Science Research Council

MAKING THE MOST OF MATERIALS

Materials and energy conservation in the materials processing industries

First published 1979 by The Science Research Council, North Star Avenue, PO Box 18, Swindon, Wiltshire SN2 1ET

ISBN: 0 900193 76 X
Designed by Banks & Miles
Printed at The Compton Press Ltd, The Old Brewery, Tisbury, Wiltshire
Cover photograph courtesy of the British Steel Corporation

Contents

The contents of each chapter are listed at the start of the chapter

Preface

This paper arose from an initiative of the Processing Sub-committee of the Science Research Council's Materials Science and Technology Committee (MSTC). Soon after its formation, the Sub-committee recognised that the conservation of energy and materials required special action. More recent developments have highlighted the national need for a co-ordinated programme of research in materials and energy conservation in the materials processing industries. The paper provides the industrial background and points to possible university research topics. It was prepared for the Processing Sub-committee by a Working Group comprising:

Dr G L Kington	Chairman (British Aluminium Co. Ltd)
Dr R D Campany	(Cambridge University)
Dr P F Chapman	(Open University)
Mr J Crowther	(British Non-Ferrous Metals Technology Centre)
Dr F Fitzgerald	(British Steel Corporation)
Mr P D R Rice	(Polymer Engineering Directorate, SRC)
Ms J E Nunn-Price	(MSTC Secretariat, SRC)

The Working Group enlisted the aid of the research associations and other experts in identifying the research needs of the materials processing industries. In particular, the SRC wishes to acknowledge the considerable assistance given to the Working Group by the following individuals and organisations:

Mr B Attwood	St Anne's Board Mill Co. Ltd
Prof J E Bailey	Redland Technology Ltd
Dr J Butterworth	Energy Technology Support Unit, Harwell
Mr H Charnock	Pilkington Bros Ltd
Dr W M Currie	Energy Technology Support Unit, Harwell
Mr P Davis	Blue Circle Industries Ltd
Mr A E Eagles	Rubber and Plastics Research Association
Dr E Emley	British Aluminium Co. Ltd
Mr A Green	Tube Investments Ltd
Dr M E Henstock	Nottingham University
Dr L M T Hopkin	Department of Industry
Mr J R Monson	British Steel Corporation
Dr R B Nicholson	INCO Europe Ltd
Dr D Pashley	Tube Investments Ltd
Mr G Perry	British Glass Industry Research Association
Mr P Popper	British Ceramic Research Association
Prof F Roberts	Consultant
Dr J A Stringer	Blue Circle Industries Ltd
Mr J Waddington	British Aluminium Co. Ltd
Dr D Watson	Blue Circle Industries Ltd
Dr D White	Leeds University

The Council hopes that this paper will be useful to research workers who are anxious that their work should be relevant to practical problems, and to industrialists who wish to encourage research in universities and polytechnics. It is therefore being published so that it can be widely read and commented upon.

In order to encourage worthwhile proposals for research into materials and energy conservation in the processing industries, the Engineering Board of the Science Research Council is setting aside an assured minimum level of funding for this research, rising from an initial annual expenditure of £50 000 in 1978/79 to £250 000 in 1982/83. These figures are intended to fund approved research programmes which fall within the materials science and technology field. The Engineering Board intends to appoint a Programme Co-ordinator who will be responsible not only for co-ordinating SRC activities, for example post-graduate training, and SRC-sponsored research programmes but also for directing potential researchers to the most appropriate funding body, for example, the Science Research Council, Department of Industry and Department of Energy. Applicants for research support are therefore advised to write, in the first instance, to

Materials and Energy Conservation Co-ordinator
c/o The Secretary
Materials Science and Technology Committee
Science Research Council
PO Box 18
Swindon
Wiltshire
SN2 1ET

who will also welcome comments.

1 Introduction

The Engineering Board of the Science Research Council has recognised for some time that it will have to be more selective in the engineering research supported by Council funds, both because of growing financial pressures and constraints and because of the need to support activities which can be shown to be important in the national interest. In fact, this development does not introduce any new principle since the SRC has followed a policy of selectivity and concentration over the past seven years. The need for such a policy, however, and for more systematic arrangements to promote it within the Engineering Board, has now intensified with the changing financial scene.

As a first step, the subject committees of the Engineering Board have defined a number of priority areas for selective support. Selective support is normally given only where the following conditions are satisfied:

1. **Excellence of study field:** Where the programme would involve work of high intellectual value, leading to advancement of the subject.

2. **Excellence of the research workers:** Where the programme will benefit from the exceptional quality of the individuals or teams to be employed, especially when there is evidence that the acknowledged leaders in the discipline have helped to formulate the programme or are eager to participate in it.

3. **Industrial, commercial and social importance:** Where the activities would improve UK manufacturing or commercial performance or the UK balance of payments, or would contribute to social aims such as environmental protection or safety at work.

The Materials Science and Technology Committee has identified the topic 'materials and energy conservation in the materials processing industries' as a priority area and wishes to promote research in this field. The national need for such research is evident from the continuing debate on energy usage, alternative energy techniques and conservation of energy and resources. The SRC wishes to encourage research into more efficient production routes and more efficient use of existing materials as well as research into new materials, the introduction of which could reduce both energy usage and the dependence on potentially scarce raw materials. For example, the Council would welcome a survey of materials already in use (possibly carried out in collaboration with the Social Science Research Council) which could point the way to energy and materials conservation. Other examples of ways in which economies in usage of energy and materials can be achieved are:

- Significantly decreasing the total energy input at key energy-intensive stages in processing;
- choosing materials for a particular use on the basis of a total energy concept;
- shifting the balance of research and development in processing to those materials which can be recycled;
- increasing the yields at each stage in a fabricating or processing chain.

The Council believes that the success of this initiative will be markedly dependent on a more closely integrated partnership between university or polytechnic departments and industry and wider application of an engineering philosophy in universities. University departments should concentrate on their traditional role of developing understanding but shift the emphasis to areas of national need, and they should make determined attempts to apply a number of principles to the solution of a specific problem, which, in many cases, means adopting a multi-disciplinary approach.

These criteria imply that SRC wishes to give strong encouragement to university groups willing to submit multi-disciplinary applications based on a close association with a processing industry.

The chapters which follow are the outcome of the deliberations of the Working Group of the Processing Sub-committee. The Working Group studied the energy and materials conservation problems of a selection of the processing industries, enlisted the aid of the research associations and other experts to identify the research needs of the industries, and identified a number of topics suitable for university research.

Because of the time and expertise available, it was necessary to limit the survey to engineering materials, so that such industries as food processing and electronic materials were omitted. It was also necessary to restrict consideration to the mainstream processing industries and this meant that the availability of the raw materials themselves (security of supply) and, at the other end of the spectrum, the total energy concept (for example, substitution technology or the cost of recovery of glass bottles) were scarcely touched upon. Within these constraints the paper represents a broad and fairly detailed view of the research needed for the immediate as well as the long-term problems arising in the industries covered by the survey.

University groups should not be deterred from proposing relevant research in areas not covered by this survey. The important factor is not so much the particular industry, as the relevance of the proposed research to energy or materials conservation within the selected industry.

In Chapter 2 the findings and conclusions are summarised and attention drawn to some of the features common to the industries studied. The rest of the paper is based on reports presented by industrial specialists, detailing present practices and current problems and highlighting specific areas in which university research could lead to significant improvements in energy and materials conservation within the industry.

An appendix describes UK Government involvement in committees, groups, schemes and research to improve energy conservation in industry.

2 Summary and conclusions

Chapter contents

Summary

In the broadest sense, this paper has its origins in social and economic pressures to conserve natural resources. In order to give useful specific guidance to universities and polytechnics it concentrates on one aspect of the conservation theme—the conservation of materials and energy in the materials processing industries. The perspective is further narrowed because, in this context, much materials saving can be related to energy saving, although there are cases where one or the other objective, but not both, can be achieved. Thus, although one chapter on recycling is included, the paper is mainly concerned with the conservation of energy, either directly or indirectly.

It is implicit in the following chapters that the necessary technological advances can be implemented only within the constraints imposed by the availability of capital, external environmental protection and the need to ensure a healthy working atmosphere. The consequence of these financial requirements and social obligations is that purely technical solutions in isolation will have little chance of being applied. This important message has been admirably and succinctly summarised by Professor H H Kellogg in the thirteenth Sir Julius Wernher Memorial Lecture in 1977: 'metallurgical processes for the future must be designed to achieve maximum conservation of energy resources, capital resources and the environment'.

This technological study of selected components of UK processing industries concentrates on the production of some important engineering materials: iron and steel, non-ferrous metals, rubber and plastics, cement, glass, ceramics and refractories, and paper and board. Each chapter reviews the energy and materials usage in the main stages of its industry and indicates the recent technological changes that have already led to improved energy conservation, either by direct saving or indirectly by reducing the amount of material processed per tonne of output. Having established the current position in an industry, an outline is then given of technological subjects for research and development that might be expected to affect the efficiency of production over about the next two decades.

The report does not set out to define complete programmes of research and development for each industrial sector, but it has attempted to emphasise and briefly describe the areas of greatest concern to UK industry and to relate them to opportunities for fruitful investigation by universities. Although this initiative originated in the Materials Science and Technology Committee, the problems go far beyond the traditional work of departments of materials science and it is hoped that the paper will provoke and guide interest in a wide range of university departments. Some typical examples of disciplines and the related areas of industrial interest mentioned in the report are given below.

Engineering design	Effects of wear of machinery on energy and material efficiency Design of better comminution machines

Instrumentation and control	Control systems for continuous casting Better instrumentation of aluminium reduction cells
Production engineering	Study of logistics of hot metal and product flow in iron and steel industry
Thermal engineering	Analysis of basic factors in heat transfer and melting Waste heat recovery
Chemical engineering and technology	Gas/powder heat exchangers Continuous pressure filters Mechanical drying
Metallurgy	Basics of continuous casting Alloys for minimal thermal treatment Upgrading and use of recovered metals
Physical and inorganic chemistry	High temperature reactions in primary and secondary smelting
Electrochemistry	Electrode processes
Mathematics	Development of models in order to understand and control processes
Physics/electrical engineering	Electromagnetics applied to liquid metals
Socio-economics	Developing total energy/economic cases for specific materials (i.e. extraction, processing, life, reclamation, re-use)

Some of the problems in the wide spectrum discussed in this report can be approached on a purely scientific basis by acquiring and analysing basic data and establishing principles (and are therefore appropriate to a specific university department and a single discipline method of thinking). The successful pursuit of many important areas, however, requires the application of an engineering philosophy—the bringing together of generalisations and disciplines to solve problems or to create new advanced processes or to create better products at lower cost. The chances of success will then obviously be increased if university departments are willing to engage in co-operative efforts. Even if the work remains on a purely scientific plane, there is still benefit to be gained from cross fertilisation. As Arthur Koestler has put it: new ideas often spring from the fusion of concepts from different fields of study.

An analysis of the wide range of subjects outlined in the body of this report shows a preponderance of projects that are specific to particular industries, but it is possible to recognise some broad common requirements, for example:

- collecting and economically using large quantities of low grade heat in cooling water and air/gas extraction systems;
- recovering waste heat by pre-heating incoming materials or by the use of waste heat boilers;
- optimising recuperators;
- improving furnace design for melting and holding metals, for example, continuous melting;
- improving refractory materials;
- moving large volumes of air at low cost for environmental reasons;
- developing better instrumentation, improving data collection, and developing process analysis and modelling techniques to improve process control;
- developing automatic control systems;
- generating a better understanding and better techniques for control in continuous casting of metals;
- developing alloys and processes to minimise subsequent thermal treatments;
- developing total-saving concepts, for example, powder metallurgy;
- processing to minimise in-house scrap production;
- developing better methods of sorting metals;
- developing better methods for removing tramp elements from metals, preferably by useful recovery;
- developing more rapid chemical reactions in secondary refining of metals;
- studying acceptable specifications and developing suitable uses for recovered materials;
- developing improved non-evaporative mechanical techniques for removing water.

There are numerous basic problems to be solved in these areas and specific questions can be posed which lend themselves to university study. Closer definition of these requirements and the necessary attention to the economic factors implicit in process improvements (or in proposals for radically new processes) can only come from an interplay between the academic and the industrial viewpoints. Consultation between universities and industry is therefore essential for the success of this programme. These discussions should be initiated at an early stage in the preparation of proposals for submission to SRC. It is also considered desirable to extend the consultative process throughout the life of the projects so that they become co-operative ventures, so helping to ensure that continuing guidance on economics and industrial relevance is available and that the results of academic work are translated as rapidly as possible into industrial practices.

Conclusions

For the reader's convenience, the industrial needs and problems identified in the main body of the paper as suitable topics for university work are collected together in the following notes.

Iron and steel (Chapter 3, page 31)

General areas

- Determine the physical and chemical properties of metal, alloy and slag melts and their behaviour in iron and steel manufacture.
- Develop mathematical models of the main extractive and refining processes in order to understand further the process in question and to develop a strategy for its control. The main processes worthy of study are the blast furnace, the basic oxygen steel-making process (BOS) and continuous casting. Further development of models of the conventional reheating furnace is not justified but work on practical convective furnaces based on the Chilton/Colburn analogy would be helpful.
- Develop techniques to use, or upgrade, low grade heat. Energy symbiosis schemes should also be examined and brought forward. Similar schemes to upgrade raw materials, recycle wastes and reduce raw material demand and processing costs are desirable.

Specific areas

- Produce better coke by use of chemical additions or treatments in non-coking blends. Partial briquetting of the charge might also be examined.
- Develop processes for enriching blast furnace and BOS waste gases, for example, on the lines of the COSORB process.
- Develop coking energy symbiosis, for example, by dry quenching coke and using the recovered heat for coal pre-heating. Similar work could be done on sinter cooling and pre-heating the air for combustion.
- Develop a method of recovering the high grade waste heat in blast furnace and steelmaking slags.
- Develop a direct reduction process which will use British coals directly (and overcome the disadvantage of high ash and sulphur content).
- Develop a more viable system for collection and enrichment of BOS gases.
- Develop a process for continuous pre-heating of charge materials to the electric arc furnace using the offtake gases.
- Use the products of direct reduction and hydro-metallurgical extraction in direct compaction (powder)/precipitation (gas) in desired product form.
- Study the logistics of hot metal and product flow with the aim of reducing energy consumption.
- Study the design of new recuperators and regenerators for waste heat recovery.
- Examine low energy gas-cleaning systems.
- Examine the recovery of scrap and scrap upgrading to remove copper and tin. Study de-tinning of tin cans and the possibility of replacing the tin by other materials.
- Examine chemical extraction of tin and/or copper from a liquid steel melt.
- Examine the practical use of indigenous iron ores. There is no possibility of physically upgrading.
- Seek to replace those metallic elements in steels which are in short supply by an examination of alloy properties and product design.
- Suggest other uses for slags from iron and steel making processes.

Aluminium (Chapter 4, page 60)

Reduction

- Investigation of mechanisms leading to current inefficiency in the Hall cell.
- Anode reactions, for example, basic investigation of over-voltage.
- Low resistance electrolytes.
- Measurement and prediction of the movement of metal layers in high currents and high magnetic fields.
- Improved monitoring equipment for measuring alumina content, temperature, and inter-electrode distance—all under working cell conditions.
- Recovering waste heat.
- Radically new process concepts, for example, permanent non-consumable anodes.
- Development of a practical and economic electrothermal reduction process.

Casting and working

- Improved method for melting and holding aluminium compared with the existing non-recuperative reverberatory furnace—preferably incorporating continuous charge feeding with counter flow of combustion gases.
- Study of the basics of containing molten aluminium by an electromagnetic field (mouldless casting).
- Alloy development for minimum energy input in homogenisation and annealing.
- Development or optimisation of low-alloyed, non-heat treatable composition for casting processes involving very quick solidification, for example, continuous casting of thin strip (6 mm) through moving moulds.
- DC casting of strip in heat treatable alloys and other compositions of wider solidification range through stationary moulds.
- Study of factors influencing the oxidation of aluminium and the growth of skim in melting, alloying and transfer operations.
- Improved methods of recovering metal from furnace skim.

Secondary smelting

- Methods for upgrading scrap to primary quality metal (particularly silicon and iron removal).
- Improved methods of sorting scrap.

Copper (Chapter 4, page 61)

Secondary refining

- More rapid reaction to shorten cycle times.
- Attention to useful recovery of other metals present as alloy or impurity.

Melting and other heating operations

- General improvement in use of furnaces by
  - basic design,
  - better operation,
  - use of waste heat.

New routes
- Development of continuous melting furnaces of wider application, with analysis of basic factors in heat transfer and in melting behaviour.
- Continuous casting of simple shapes.
- Production of forms other than cathode plate in electrorefining.
- Production of strip from powders.

Recycling
- Better methods of identification and separation of alloy types.

Zinc (Chapter 4, page 62)
Melting and casting
- Better design and use of furnaces in melting.
- Better throughput and product yield in the use of pressure die casting machines, mainly by more instrumentation and reactive control and more thorough analysis of the design of castings.

Recycling
- Increased recovery of zinc from the fumes arising from zinc-coated steel scrap and brass melting (containing zinc oxide), which at present are not economically treatable.

Lead (Chapter 4, page 62)
Refining
- Reduction of energy demand by more rapid reactions and more efficient furnaces.

Nickel (Chapter 4, page 62)
Recycling
- Improvement in recycling of plating solutions.
- Better methods of identification and sorting of solid articles in various nickel-base materials.
- Improvement of solvent extraction techniques for secondary or primary nickel in solution.

Titanium (Chapter 4, page 62)
- Making castings direct from the melting process.
- Casting a rolling slab.
- Casting a small diameter bar in long lengths.
- Continuous casting of shapes for working.
  (In all these there are basic problems in linking the casting stage to the continuous vacuum melting and, in the last at least, in matching melt rate and freezing rate.)
- More efficient methods of pre-heating for working (may be by induction heating).

Magnesium (Chapter 4, page 62)
- Method of cleaning liquid metal without use of salt fluxes.

General non-ferrous metals (Chapter 4, page 62)
In all melting and heating processes, particular attention needs to be given to the recovery of waste heat, preferably by pre-heating incoming materials. The recovery of low-grade heat contained in gases and cooling waters is a general problem.

Rubber and Plastics (Chapter 5, page 90)
Conservation of energy
- Theoretical energy requirements of major industrial processes. To extend over the production of polymers and major fillers from basic raw materials, mixing and compounding procedures, and major forming processes (for example, calendering, extrusion, compression moulding, transfer moulding, injection moulding, blow moulding, vacuum forming and vulcanisation techniques).
- Better techniques for the recovery of waste heat particularly when contaminated. To consider air extraction systems for vulcanisation equipment, reinforced plastics lay up, resin impregnation processes, solvent based spreading techniques, and drying processes. To consider condensate return systems for chemical reactors (where potential contamination by leakage is present), autoclave vulcanisation, and processes using live steam injection.
- Better techniques for the upgrading of waste heat. To use the large volumes of low grade heat lost in cooling and ventilation systems.
- Improved efficiency of energy use in providing drives. To study the significant energy losses associated with large drives required to meet intermittent high loads in processing machinery and with services such as compressed air.

Conservation of both materials and energy
- Theoretical effects of changes such as wear of machinery on the energy and material efficiency of major processes. To investigate the loss in effectiveness due to alterations in clearances in such equipment as internal mixers, extruders and injection moulding machines. Data are required to establish the time for machinery reconditioning or replacement for energy and material optimisation.
- Study of acceptable levels of diversity in recovered polymeric materials. To measure levels of dilution by, for example, paper and the extent to which mixtures of polymers can be tolerated. This is important to the levels of separation that are required of collected waste.
- Recovery of materials such as solvents, monomers, etc, in total loss extraction systems and the recovery and recycling of heat contained in such systems.

Glass (Chapter 6, page 106)
- Develop useful ways of conserving the considerable amounts of low-grade heat (for example, cooling water at 50°C) rejected by glass-making processes.

- Develop higher quality refractory materials for furnace superstructure, regenerators and glass contact to improve both fuel utilisation and glass quality.
- Improve understanding of the basic melting and refining operations in order to design more efficient furnaces and control them more closely. The ability to predict flow and temperature regimes within the glass in novel designs of melter would be desirable. Present studies concentrate on modelling existing systems.
- Examine low temperature routes to glass product manufacture.
- Investigate the possibility of producing glass directly from a solid fuel energy source.

Ceramics (Chapter 6, page 107)
High temperature processes
- Study the thermal behaviour of kiln walls with various types of construction.
- Match kiln wall structure to the firing cycle and to the production schedules of specific products.
- Investigate the stability of fibrous insulation materials in various environments.
- Study fast firing techniques.
- Examine the changes needed to body compositions and raw materials to permit lower firing temperatures.

Substitute materials
- Extend the engineering uses of the newer ceramics.
- Investigate systematically the development of low temperature bonds and desired microstructures in ceramic materials.
- Investigate the fabrication of ceramic materials suitable for making large sheets.
- Develop new and cheap ceramic coatings and coating methods.

Cement (Chapter 7, page 139)
Kiln process
- Design of efficient and compact cyclones for gas-powder heat exchange.
- Prediction and control of ash deposition from the flames of high ash fuels.
- Design of pulverised coal firing pipe injectors to reduce the required injection velocity, fanpower and fan wear.
- Improved refractories (better life, lower thermal conductivity, lower cost).
- Uses for large quantities of low grade heat.
- Studies of gas and solids flow and heat transfer in furnaces.

Comminution
- Design of more efficient machines capable of scale-up to 100 tonne/hr or more, with acceptable repairs costs when grinding abrasive feeds.
- Better understanding of grinding characteristics of materials.
- Better understanding of the mechanism of grinding aids.

Chemistry

- A chemical approach to lower energy requirements, particularly energy for cement clinker grinding.

Rheology

- Additives to maintain the flow of cement feed slurries with low moisture content, particularly in the presence of chloride and sulphate ions.
- Design basis for a truly continuous pressure filter.

Paper and board (Chapter 8, page 155)

- Improvement of refining techniques and control. The efficiency of the present refining process is low, mainly because of the difficulties encountered in effectively transferring energy to the paper fibres in order to achieve the required modifications. The use of ultrasonics may provide one approach to the solution, but there could be others.
- Investigation of the chemical preparation of secondary fibres and the effects of repeated recycling on fibre properties.
- Examination of substitute materials for wood pulp, such as straw, willow and synthetic materials.
- Further research into head box design and its influence on sheet formation. Including an examination of the effects of an increase in stock consistency on energy use.
- Improvement of mechanical drying techniques through felt development, and the enhancement of press action by pre-heating, etc.
- Further study of the combination of water with the fibres in the web and the changes that occur during drying, with particular reference to the identification of important variables and the improvement of current drying performance in the final stages.
- The development of product drying controls and monitoring systems for all stages of processing.
- Further research into dry forming techniques and the associated problems of stock preparation and final product strength.

Reclamation and recycling (Chapter 9, page 170)

Technological and techno-economic

- Many proposed methods of energy recovery are at the stage when only extensive data from pilot plants would be of use. Hydrolysis of cellulose in domestic refuse (to produce sugars) and of non-metallic residues from scrap shredders is worth further scrutiny on a laboratory scale.
- Assessment of the short- and long-term effects of the build-up of contamination in materials, such as copper and tin in steel or lead and aluminium in zinc.
- Reassessment of composition specifications for materials.
- Development of rapid methods of materials identification at the scrapyard.
- Accurate measurement of melting losses and energy requirement for reclamation from different types of arising in order to determine the optimal recycling level for each type.

- Analysis of the true level of materials costs in manufacture in order to establish the true benefits of materials substitution.
- Examination of the costs of design changes to facilitate recycling of manufactured goods, for example, cars and other complex structures.
- Development of open-loop recycling where degraded products with low value are used in applications where purity is not important.

Economic and socio-economic

- Assessment of the true economics of collection (especially by volunteer groups) of paper, beverage containers, etc.
- Stabilization of supply and demand relationships for secondary materials such as paper or steel.
- Development of a systems approach so that accurate predictions can be made of all the effects of the recovery and re-use of industrial products or materials. The glass container industry is a heavy consumer of energy and might be a particularly useful subject for study.

3 Iron and steel

Chapter contents

Introduction

It is essential to consider energy and materials conservation together in the iron- and steel-making industries since loss of partially or fully processed material represents a loss of the energy which had already been expended on the material. Furthermore, the steel industry makes considerable use of recycled materials in the form of scrap, therefore reducing the energy consumption which is necessary to separate virgin iron units from the oxygen with which they are invariably associated in nature. Materials conservation in the sense of optimising designs and developing high strength steels to minimise the quantity of steel required for a particular purpose will also tend to reduce overall energy consumption by the industry.

Total energy demand

The manufacture of steel is an energy intensive industry. In 1975, the industry in the UK consumed approximately 507 million GJ or 25·1 GJ/tonne of crude steel produced. A high proportion of this energy is rejected, much of it in the form of warm water. The total energy usefully employed is only about 30%, made up of the energy to reduce iron ore plus the energy sold in by-products, although a higher figure is more realistic to take account of the intermediate heating operations necessary to convert iron to finished products. It is estimated that 29% of the energy input is rejected as waste gas and cooling water and a further 17% is lost by furnace radiation.

About 60% of the energy input is in the form of coal, most of it being used for the manufacture of coke, and 20% is oil which is mainly used for steam raising and power generation.

Energy used in iron-making

The role of coke

Almost all the energy consumed in the blast furnace/BOF route up to the ingot stage is required for the production of iron, the principal energy source being coal which is converted to coke. There is also significant production of energy in the form of coke oven and blast furnace gas, which is utilised in the finishing stages.

The coke plays both chemical and physical roles in the operations as distinct from just being a source of energy. In order to produce coke of suitable quality, it is essential to have a good coking coal, but unfortunately the UK is now short of good coking coals and relies on significant imports. Not only are the UK reserves of good coking quality being exhausted, but such as remain have a high sulphur content.

In order to produce low sulphur iron from high sulphur coal, large quantities of limestone must be used. Extra energy is then needed to decompose the carbonates and slag production is increased, adding to the heat losses. An alternative approach is to desulphurise the liquid iron prior to refining in the basic oxygen furnace and this practice is being widely adopted, but it is time consuming, expensive and involves a slight increase in energy consumption.

The iron and steel industry is actively engaged on studies to reduce the coke consumption of the blast furnace and to make an acceptable coke from British coals.

To illustrate these studies, it is necessary to examine the major roles of coke in the iron blast furnace. First, the coke supports the burden and maintains the permeability of the column to the hot blast. If the coke were to degrade, the permeability of the burden would be decreased, the flow of hot blast would be reduced and the production from the furnace would similarly be reduced. Second, the coke supplies carbon and carbon monoxide and, third, it supplies heat in the hearth of the furnace for melting. The carbon monoxide, generated by reaction of coke with the hot air blast in the hearth of the furnace, is used for reduction of the ore in the bosh and stack, when it is oxidised to carbon dioxide. In the higher temperature regions (the bosh) the carbon dioxide is immediately reconverted to carbon monoxide by reaction with some of the coke (carbon solution loss). In the hearth of the furnace, melting and separation of metal and gangue (i.e. slag) is achieved.

Coke demand for the melting function can be reduced by hydrocarbon injection at the hearth through tuyeres. Coal, oil or natural gas are already being used for this practice. Coke savings of about 150 kg/tonne of hot metal produced are possible, the coke rate dropping from about 520 kg/tonne to 370 kg/tonne. As an alternative coke economy, reducing gases may be introduced in the bosh of the blast furnace in order to overcome the carbon solution loss in the stack. Coke savings as high as 150 kg/tonne have been postulated theoretically for this procedure but the scheme has not yet been carried out on a practical commercial scale. The effects of fuel injection at the tuyeres and of bosh injection are largely additive and coke savings of the order of 270 kg/tonne of hot metal produced could be achieved. Use of oxygen fed burners instead of hot blast tuyeres (i.e. nitrogen-free operation) would result in the production of a high calorific value top gas which could be treated and recycled, giving a very low coke rate (as little as one third of that achieved in normal operation). Such a practice has not yet been tested on a full-scale furnace.

In discussing these practices, it is necessary to appreciate that they would have a far reaching effect on the net energy requirements in an integrated works since the blast furnace is a producer of fuel gases as well as metal. Its demand for coke governs the availability of coke oven gas for fuel and coke breeze for sinter manufacture. The overall effect of coke economies in practice could only be assessed by use of a fully integrated strategy model for works' energy demand and production.

Although these modifications to blast furnace practice reduce coke requirements, it is necessary to examine all the processes which might extend the strained coking coal reserves of the UK. Coke properties can be improved by pre-heating the coal before charging it to the coke oven. The coal is pre-heated to about 200 to 260°C, then charged to the oven through a pipeline in a steam flow. Marginally better coke is produced from the same blends and coke oven productivity is improved by up to 40%. It is also possible to use coals not normally used for coke production by using the "plum duff" concept, i.e. partial briquetting of the non-coking portion of the coal blend prior to carbonisation. This technique is already practised in Japan but is not yet proven for British coals. Hot formed briquettes, sometimes subsequently carbonised and termed formed coke, are also receiving attention.

Energy recovery in conventional iron-making

Potential areas for improved energy recovery in conventional iron-making plants are generally limited to recuperation of sensible heat, although the quality of both coke oven gas and blast furnace gas, which are already collected and utilised, can possibly be improved. The low calorific value of blast furnace gas, mainly due to its high nitrogen content, reduces its utilisation efficiency. Some of the modifications to blast furnace practice outlined above will help to correct this but consideration could also be given to upgrading the gas to produce a richer fuel or even a chemical feedstock. Techniques for doing this are known but are capital intensive. Care must be taken that the energy involved in such up-grading does not exceed the losses experienced in conventional use. The most immediate improvement which could be made in utilisation of these arising gases is better energy management within integrated works to minimise gas flaring.

The sensible heat of coke as it is pushed from the coke oven is generally wasted. This amounts to about 1 500 MJ/tonne of coke. Gas cooling techniques have been developed to replace the conventional water quench but the heat thus recovered is generally used for steam raising and does not represent an attractive investment at most UK steelworks. Alternative methods of using the heat, such as applying it to pre-heat the coal, need to be developed.

Significant sensible heat losses occur in the ore sintering process, both in the waste gases and the hot sinter. Because of the large volumes concerned, the gas losses are high, although average temperature is low (100 to 200°C). The sensible heat of the sinter, however, could be recovered and used to pre-heat the incoming combustion air, effecting savings of up to 300 MJ/tonne of sinter.

Energy losses in the blast furnace are principally in the form of cooling water and molten slag. The temperature of the cooling water is only raised to 30 to 60°C (unless evaporative or steam raising cooling is employed) and therefore has limited application although it amounts to about 400 MJ/tonne of iron. The heat of the molten slag is similar in magnitude but at high temperature (1 400°C). As yet there is no satisfactory technology for recovering this heat.

As high top pressure becomes more widely used on UK blast furnaces, an opportunity will arise for recovering some of the gas pressure energy by means of expansion turbines. This can help to offset the increased energy required by the blast furnace air blowers in such operation. About 25% of the blower energy is recoverable in this way.

Alternative methods of iron-making

Coke free

Coke can be eliminated completely from iron-making in the direct reduction processes. The Midrex process is the basis of two such plants chosen for Hunterston.

These Hunterston plants are based on natural gas, but this will not always be readily available in the UK. There is therefore a need to have a practical coal based process by the turn of the century. Incorporation of a step to convert the coal to a gas could extend the use of Midrex type processes but the direct use of coal would simplify matters. Two such processes are of interest and do not demand coking coal. The first is the SL/RN process and the second is the centrifugal iron-making process. The SL/RN process has not operated efficiently using British coals (with their high ash and sulphur content) and the other is only at the pilot plant stage. It would require both scaling up and a good waste gas heat recovery system to be effective. Unlike other direct reduction processes, however, it produces a liquid metal and therefore does not require further energy in the form of electricity for melting.

A third alternative, which uses coal as a reducing agent and other fuel as a heat source, is the Kinglor Metor process. In this process, the iron ore and solid reducing agent are passed continuously through a vertical ceramic tube, which is externally fired. The Kinglor Metor process is in the early stages of industrial application in Italy, but has not yet been tested on a wide variety of materials. In particular, it has not been tested with UK coals.

Cupolas

Iron is also melted in cupolas, which are associated with foundries making iron castings. In 1975, about 3 million tonnes were produced in this way. The cupola is a shaft furnace in which iron scrap, steel scrap and some cold blast furnace pig iron is melted by burning coke (which is charged with the scrap) with air blown through tuyeres. The principle is similar to a blast furnace but units are very much smaller and coke requirements very much less because there is no reduction carried out and little gangue to be removed as slag. In an attempt to completely remove the necessity for coke, a gas- or oil-fired cupola has been developed to the trials stage. Some cupolas are being replaced by electrical induction furnaces, which will also eliminate the need for coke.

Energy used in steel-making

The Basic oxygen furnace (BOF)

In the BOF steel-making process, energy is required to melt the 20 to 30% scrap addition which is normally made. This energy is provided by oxidation of the carbon and silicon contained in the blast furnace hot metal.

Energy is also obtained by oxidation of a small proportion of the iron and from excess sensible heat in the blast furnace hot metal. When considering this process it should not be forgotten that an energy input is also required to produce the oxygen used.

A gas rich in carbon monoxide is evolved during the steel-making process. In the majority of UK plants this is quenched and flared with the loss of about 800 MJ/tonne of liquid steel. Two methods of recovery are available. In the first, the gases are burned at the vessel exit and the heat used to raise steam in waste heat boilers. Up to 75% of the heat can be recovered in this way but engineering and maintenance problems are severe. There is a world-wide tendency to abandon this method in favour of a suppressed combustion system in which the gases are collected and distributed to other parts of the works for use as fuel. About 400 MJ/tonne of steel can be recovered in this way, the gas having a calorific value of 6 600 KJ/Nm^3. This gas, like blast furnace gas, could also provide the basis for a chemical feedstock. It would be less difficult to use than blast furnace gas as it contains about 90% carbon monoxide and would not require the handling of large volumes of unwanted gases such as nitrogen.

The sensible heat of molten basic oxygen steel-making (BOS) slag accounts for an energy rejection of about 250 MJ/tonne of liquid steel. The recovery problems are similar to those involving blast furnace slag and no technical solution has yet been found.

Other methods of conversion

Some steel is still produced in open hearth furnaces, which are usually oil-fired. This process can accept a much wider range of scrap to hot metal ratio but is more energy intensive than the BOS process. Appreciable opportunities for energy recovery exist and some were developed over the many years when this was the major steel-making process. The application of waste heat boilers is the most notable of these. The process is now in decline, however, and further efforts to improve its fuel economy are not justified.

The major alternative to the BOF process is electric arc steel-making using charges composed principally of scrap metal. This is also the major outlet for the products of direct reduction plants. In 1975, 5·5 million tonnes of steel were made in UK arc furnaces. A typical figure for the energy used is 2 300 MJ/tonne of liquid steel produced of which 1 660 MJ/tonne is provided by electricity and the remainder by chemical reactions in the steel. The bulk of the energy input leaves the furnace in the liquid steel (60% or 1 380 MJ/tonne). The remainder is lost in the slag (5%), in the waste gases (14%), in various water cooling systems (5%), through electrical losses (8%) and other losses including those through the furnace structure (8%).

The waste gas is at a temperature of 1 000 to 1 600°C and contains appreciable quantities of combustible gases. This represents the only potential high grade waste heat source except for the slag (which is only an intermittent product). Research into the utilisation of the waste gas as a heat source could be of benefit, but the possibilities are limited by the rapidly varying nature of the process. The feasibility of upgrading the low grade waste heat sources by, for example, replacing water cooling systems by evaporative cooling ones, is of some interest.

Scrap pre-heating can reduce the electrical energy consumption of the arc furnace and when the furnace is close to a direct reduction plant the possibility exists to hot charge the direct reduction material. This would reduce electrical consumption and eliminate the necessity for a cooling section at the direct reduction plant.

Energy used in finishing

Much work has gone into studies of how the considerable amounts of energy contained in newly refined molten steel might be best utilised in the casting and soaking of ingots. Within the metallurgical and practical constraints, every effort is made to charge ingots to soaking pits as hot as possible. Despite these efforts, up to 12% of the energy used by the steel industry is spent on re-heating in the forming processes. There are numerous ways of reducing this—some of them are developed but others require further work before commercial application is possible.

Reducing the number of processes

One method is to reduce the number of forming processes and hence the number of re-heating processes. The most notable technique in this area is continuous casting, whereby the ingot stage and soaking pits are bypassed and liquid steel is formed directly into slabs or blooms. The liquid steel has a sensible heat of about 1 400 MJ/tonne and much of this is rejected in the casting mould cooling water at a temperature of 30 to 40°C.

Powder forming

The ultimate stage of minimising re-heating processes is achieved in powder metallurgy routes in which fine metal particles are formed directly into the desired finished shape. This method has not proved suitable for anything other than very small components so far, but development work is continuing to apply it to more general products. The metal powder can be produced either by the atomisation of liquid steel or by direct reduction of very pure iron ores. The latter technique is more attractive from the energy point of view since it avoids the very high temperatures associated with a liquid route and is potentially a better economic proposition. If the engineering problems associated with these techniques can be overcome, it also opens up the possibility of extracting iron powders from impure ores by low temperature chemical routes. These have been examined on numerous occasions but have yet to prove themselves suitable for commercial production of steel in large quantities. Apart from potential energy savings, however, the overall improvement of yield which would result from the elimination of forming and machining losses makes further effort attractive.

Continuity between stages

In the conventional forming routes, the attainment of more continuity between stages offers potential energy savings although the need for surface treatment between stages makes this difficult.

The development of instrumentation for on-line assessment of hot slabs, for example, from either a continuous slab caster or a slabbing mill would lead to the use of automatic hot scarfing machines and enable more continuous processing to be carried out. Even without such physical continuity there is scope for reduction of re-heating energy consumption by better scheduling of material through the various process stages, minimising the length of time re-heating furnaces are maintained at temperature awaiting the next charge.

Heat recovery

Schemes for the recovery of heat from re-heating furnace waste gases are constantly being developed. The most recent involves the use of a ceramic recuperator but further improvement is possible. The use of evaporatively cooled skids in re-heating furnaces instead of water cooling would permit the recovery of heat lost to this source which is currently rejected as warm water. Such schemes tend to have a high capital cost however, and justification is difficult at the present time.

Other energy uses

Environmental protection

One area where energy consumption is likely to increase rather than decline during the next few years is in environmental protection, both in terms of improving atmospheric conditions in working areas and improving the cleanliness of effluents discharged from the works. An indication of the energy associated with such schemes is given by the fact that power consumption for the air cleaning system of a blast furnace cast house is similar to that required to drive the furnace blowers themselves. The development of low energy gas cleaning systems is being pursued.

Most of the energy for environmental control will be in the form of electricity, which introduces its own production losses at the generating station. Care must be taken to see that attempts to improve environmental conditions at the steelworks do not introduce an equal pollution problem at the generating station as well as increasing energy losses.

Recovery of secondary low grade energy

Much of the energy rejected throughout steel-making and finishing is low grade and in the form of warm water. Many of the potential energy recovery techniques involve steam raising, some of which might be used for electricity generation which will then reject a high proportion as its own low grade energy losses. Various schemes for utilising low grade energy more efficiently have been proposed, including district heating, horticulture and fish farming, and some of these have been put into practice on a limited scale. There would appear to be considerable scope, however, for research into methods of economically upgrading such energy. The expansion through a turbine of fluids of suitable critical temperature might permit the generation of additional electricity. Techniques such as this would be of great value not only to the steel industry but to a wide range of energy consumers.

Raw Materials

Scrap

In 1975, the steel industry consumed more than 16 million tonnes of scrap, about 4 million tonnes being used in iron foundries and most of the remainder for steel-making. Scrap represents about 50% of the iron processed into liquid steel, so saving the energy required to reduce iron ore. About 7 million tonnes of this scrap occurred as in-house arisings (i.e. re-circulation within the works) the remainder being recovered from outside the steel industry. With the pressures to conserve resources and improve the quality of the environment, it is likely that types of scrap not currently recovered will be utilised in future. Some 0·75 million tonnes of tin cans are thrown away each year and the total volume of ferrous materials in household waste is of the order of 1 million tonnes/year. There are also economic pressures to recover scrap material—the use of 1 tonne of scrap in steel-making saves about 1·7 tonnes of ore, 0·68 tonnes of coke and 0·28 tonnes of limestone. In 1975, it was estimated that the savings to the balance of payments to the country by the use of scrap was about £100 million/year.

The problem with the use of scrap is the presence of residual elements (mainly copper and tin) which generally have a harmful effect on steel quality, although for certain specialised uses about 0·5% of total residuals has a beneficial effect. In-house arising scrap has a copper content of 0·025 to 0·06% and a tin content of between 0·01 and 0·02%, compared with merchant scrap which can have levels greater than 0·3%. The average British car contains about 1% copper by weight, although by removing the engine, battery and cooling system, the copper content is reduced to about 0·6%. It is reported that the copper content of the product from a conventional fragmentizer can be as low as 0·20% and from a shredder/ripper fragmentizer (where there is less entrapment of non-ferrous material by the ferrous product) it can be as low as 0·12%. There is much work directed at reducing the entrapment of non-ferrous materials by the ferrous product, the bulk of this work being on cryogenic fragmentation—copper contents as low as 0·08% have been quoted from a Belgian plant. Cryogenic fragmentation and tin removal seem to be essential parts of a process to recover tin cans and give a product acceptable for fine steel-making.

Although the chances of success are very slim, it is worth examining the possibility of removing copper and tin from the liquid steel melt. A suitable slag or other reagent acting at liquid steel temperatures would be required.

The recycling of used ingot moulds presents special problems because of their size, particularly when they are recycled to foundry cupolas. It is necessary to break them into pieces having a maximum linear dimension of 50 cm and this is normally carried out by smashing with a steel ball. Recent developments to improve mould performance have involved the use of quasi-flake graphite iron instead of conventional flake graphite iron, but these have proved difficult to break by the drop ball method and alternative techniques are required. Cryogenic fragmentation again offers possibilities but in this case capital and energy costs would be high due to the size of the moulds involved.

Other materials

Ore

The bulk of iron ore used in the UK is imported. For strategic reasons it is essential that the efficient use of indigenous ores (with their inherent high gangue content) is technically developed. Because most of the gangue is in chemical combination with the iron, normal mineral dressing techniques are not possible and chemical treatments may be necessary. The potential for combining such techniques with the production and forming of iron powders has already been noted as a possible energy saving measure but account must be taken of the energy required both to produce any chemicals which are used and to ensure cleanliness of effluents.

Dust

About 600 000 tonnes of flue dust are produced each year, composed of iron oxide with various contaminants. Approximately 360 000 tonnes come from the blast furnace, 50 000 tonnes from the arc furnace and 160 000 tonnes from the BOS process. Much of this material is recycled but some has to be dumped. A scheme to recover both the iron and the zinc (which is combined with the arc furnace dust in particular) as prime units is under consideration but is costly in energy. Alternative methods should be studied.

Alloying elements

Many steels contain significant portions of alloying elements (for example, chromium, nickel, manganese, molybdenum, titanium, etc.), all of which will ultimately be in short supply. Alternative alloying elements are being sought as are means of improving the recovery of these elements in the steel-making process. Development of new alloy steels will reduce energy consumption in some areas by eliminating the need for hardening and tempering operations after forging.

Slag

The major solid by-product of the steel industry is slag, of which about 10 million tonnes is produced each year from blast furnaces and steel furnaces. The potential of this material as a source of high grade heat has already been mentioned. All of the blast furnace slag (about 5·6 million tonnes/year) and some of the electric arc furnace slag is used, mainly for the production of roadstone. Some of the BOS slag is recycled to the blast furnace as flux although the extent to which this is possible is limited by the build up of contaminants. Other outlets for these by-products are urgently needed.

University research

Some suggestions for university work to improve the conservation of energy and materials in iron- and steel-making are listed below.

General areas

- Determine the physical and chemical properties of metal, alloy and slag melts and their behaviour in iron and steel manufacture.
- Develop mathematical models of the main extractive and refining processes in order to understand further the process in question and to develop a strategy for its control. The main processes worthy of study are the blast furnace, the basic oxygen steel-making process (BOS) and continuous casting. Further development of models of the conventional reheating furnace is not justified but work on practical convective furnaces based on the Chilton/Colburn analogy would be helpful.
- Develop techniques to use, or upgrade, low grade heat. Energy symbiosis schemes should also be examined and brought forward. Similar schemes to upgrade raw materials, recycle wastes and reduce raw material demand and processing costs are desirable.

Specific areas

- Produce better coke by use of chemical additions or treatments in non-coking blends. Partial briquetting of the charge might also be examined.
- Develop processes for enriching blast furnace and BOS waste gases, for example, on the lines of the COSORB process.
- Develop coking energy symbiosis, e.g. by dry quenching the coke and using the recovered heat for coal pre-heating. Similar work could be done on sinter cooling and pre-heating the air for combustion.
- Develop a method of recovering the high grade waste heat in blast furnace and steelmaking slags.
- Develop a direct reduction process which will use British coals directly (i.e. overcome the disadvantage of high ash and sulphur contents).
- Develop a more viable system for collection and enrichment of BOS gases.
- Develop a process for continuous pre-heating of charge materials to the electric arc furnace using the offtake gases.
- Use the products of direct reduction and hydrometallurgical extraction in direct compaction (powder)/precipitation (gas) in desired product form.
- Study the logistics of hot metal and product flow with the aim of reducing energy consumption.
- Study the design of new recuperators and regenerators for waste heat recovery.
- Examine low energy gas-cleaning systems.
- Examine the recovery of scrap and scrap upgrading to remove copper and

tin. Study de-tinning of tin cans and the possibility of replacing the tin by other materials.

- Examine the chemical extraction of tin and/or copper from a liquid steel melt.
- Examine the practical use of indigenous iron ores. There is no possibility of physically upgrading.
- Seek to replace those metallic elements in steels which are in short supply by an examination of alloy properties and product design.
- Suggest other uses for slags from iron- and steel-making processes.

4 Non-ferrous metals

Chapter contents

Introduction

Only four non-ferrous metals, aluminium, copper, zinc and lead, are used recognisably in large quantities, and even then at levels very much lower than that of steel. The biggest use of several others is as additions to iron and steel, but the production of ferro-alloys takes place largely outside the UK and is not relevant to this review. Current consumption of all the non-ferrous metals is shown in Table 4·1. Aluminium and copper each have a wide range of uses, which provides the basis of an industrial production pattern not dissimilar from finishing in the steel industry, but the industrial treatment of lead and zinc is much more limited.

Table 4·1
Non-ferrous metal consumption in the Western World, 1976

Metal	Consumption (tonne $\times 10^3$)	Uses
Aluminium	10 500	Structural, transport (road, marine, air), electrical, packaging, holloware
Copper	6 200	Electrical, plumbing, alloys (brass, bronze, etc)
Manganese	5 000	Almost entirely as ferro-alloy, 50 000 tonne as electrolytic metal for special alloys
Zinc	4 000	Die castings, batteries, galvanising, alloys (brass, etc)
Lead	3 000	Secondary batteries, plumbing, solder, cable sheathing, radiation screens
Chromium	2 000	Almost entirely as ferro-alloy, 2 000 tonne other alloys and uses
Nickel	650	Aerospace materials, alloying (stainless steel, armour plate, coinage, etc), secondary batteries
Magnesium	250	Alloying with aluminium, chemical/electrochemical, structural (aircraft, transport, hand tools/luggage)
Tin	190	Tinplate, alloys (solder, babbit)
Titanium	50	Aerospace, heat-exchangers, chemical plant
Molybdenum	50	Mainly as ferro-alloy
Cobalt	25	Aerospace, tools
Vanadium	25	Almost entirely as ferro-alloy
Tungsten	25	Carbides, filaments, heavy-metal

[1] Although present in metallic form, electroplating does not represent a true use of chromium since the plating solution is prepared from the ore and not from the metal itself.

Primary metal is obtained from naturally occurring minerals by reduction processes which demand the expenditure of chemical energy; for most metals, this is considerably greater than the energy subsequently needed for the heating and shaping processes involved in making useful articles out of the metal. In pyrometallurgical processes the energy is usually provided by a carboniferous fuel although in favourable cases, such as copper sulphides, the burning of the sulphur itself makes a considerable contribution. In other cases, such as aluminium smelting, the reduction is accomplished by the use of a secondary form of energy, electricity. Hydrometallurgical processes may use energy indirectly (as when dissolving minerals in acids) but also call for considerable amounts of mechanical energy in moving materials. For almost all copper, half the world's zinc, and much lead, the final separation as metal is by an electrolytic process.

All the metals must be converted to liquid for alloying and transformation into useful shapes, either directly by casting in a foundry or indirectly by casting a cylindrical or prismatic shaped block for subsequent mechanical deformation. Unlike steel, the blocks are usually allowed to cool after casting and must be re-heated for hot working, although in some recently introduced processes continuous casting and rolling is carried out. Although rolling is the commonest working process, it is not so completely dominant as with steel; the extrusion of bar and shapes is important for both aluminium and copper materials whilst forging is also signficant for aluminium. Foundry casting of non-ferrous metals is relatively a bigger activity than with ferrous metals.

A quarter of UK zinc is used as die-castings, a quarter disappears into the copper industry as brass and another quarter into the steel industry as galvanised coatings. Lead still has a wide range of uses, but is increasingly dominated by lead-acid secondary cells. With both these metals, working processes play little part.

In the aluminium, copper and lead industries there has long been a well developed recycling activity.

Energy conservation and metal conservation clearly interact, particularly where metal losses are concerned. As far as possible, they are discussed in an integrated manner in the following sections.

The aluminium section is somewhat longer than the others, partly because primary smelting of aluminium is carried out in the UK (unlike copper) and partly because some topics applicable to other metals are for convenience placed in the first section.

Aluminium

Approximate data for the production and utilisation of aluminium in the UK are given in Table 4·2.

Note: Energy requirements in this chapter are expressed for convenience both in gigajoules, GJ, and megawatt hours of thermal energy, MWh(th). Electrical energy is expressed in megawatt hours, MWh(e), and may, where appropriate, be converted back to the fossil fuel equivalent expressed in GJ and MWh(th).

Table 4·2
Production and consumption of aluminium in the UK, 1976[1]

	Aluminium	Quantity (tonne × 10^3)
Primary		
	Production	335
	Consumption[2]	460
Secondary		
	Production	205
	Scrap consumption	245
Sales		
	Rolled products	215[3]
	Extrusion and tubes	140
	Castings	120
	Foil	40
	Drawn wire products	33
	Powder	8
	Forgings	4
	Paste	3
	Total	563
Imports		
	Unwrought	220
	Rolled	80
	Extruded	22
	Scrap	10
	Total	332
Exports		
	Unwrought primary	118
	Unwrought secondary	44
	Rolled	30
	Extruded and drawn	19
	Scrap	1
	Total	212

1 The table is compiled from statistics circulated by the Aluminium Federation and the Department of Trade and Industry.

2 Consumption figures represent intakes of primary metal by UK consumers and do not take into account changes in consumers' stocks.

3 Includes 60 000 tonnes for foil stock.

Smelting

The energy demand of primary aluminium production is high and can be roughly broken down as shown in Table 4·3.

The refining of bauxite to alumina is carried out mainly outside the UK but it is accepted that there is considerable scope for economy by recuperation and improved plant design.

The high energy demand of the electrolytic process has been steadily reduced over the past 25 years from 22 to 13 MWh/tonne in best practice,

Table 4·3
Average energy requirements for production of aluminium

Production stage	Energy requirement (GJ/tonne Al)	(MWh(th)/tonne Al)
Mining and shipping bauxite	Variable	Variable
Refining to alumina		
Bayer process	21	5·8
Calcination	11	3·1
Electrolysis to metal		
Electric power	50	14[1]
Carbon anode contribution	16	4·4

[1] Measured in MWh(e).

although the bulk of installations must lag behind this. The energy demand is, in effect, much greater if the electrical energy has to be produced from fossil fuels. Whereas more than half of world production is from hydro-electric power, in the UK 90% is from fossil fuels or nuclear energy and energy saving is correspondingly more important.

Efficiency of conventional electrolysis

The theoretical energy requirements for the electrolysis process are:

- using carbon electrodes, as in normal industrial practice ($2Al_2O_3 + 3C = 4Al + 3CO_2$) — 5·64 MWh(e)/tonne
- with a non-reactive electrode ($2Al_2O_3 = 4Al + 3O_2$) — 8·67 MWh(e)/tonne

The energy efficiency can best be considered as having two components, current efficiency and voltage efficiency. Over the past 25 years, current efficiency has improved from about 85 to 90%, with possible efficiency now at somewhere about 94%. The cell voltage (4·2 V on a modern cell) is roughly two and a half times the theoretical level.

Loss in current efficiency is due almost entirely to metal from the cathode diffusing back towards the anode and being reoxidised by the escaping anode gases. The exact mechanism—solution, dispersion, subvalent species, etc—is not understood; nor are the factors which influence it, with the exception of temperature. There is a need for fundamental knowledge to guide further development.

Excess voltage is due to overvoltage and to the resistance of the electrodes, leads and electrolyte layer. The overvoltage at the carbon anode is about 0·6 V. It is not known whether it might be possible to reduce this by changing the electrode process by, for example, a more reactive carbon or by supplementary reducing gas. The cathode overvoltage is small. Resistance of the electrolyte layer is a function of resistivity, working distance and current density. There has been a search for lower resistivities by the use of appropriate additives, chiefly lithium salts.

Both overvoltage and metal loss by back reaction may be functions of alumina content. Again, the knowledge is not sufficient to guide operating practice.

Also, the two components of performance, voltage and current efficiency, are not entirely separate. Interelectrode distance can be reduced to save volts but current inefficiency will increase not only because of back diffusion processes but also because of instability of the cathode (the surface of the metal pool) which may lead to direct shorting. Under the influence of the magnetic fields, circulation and wave effects occur in the metal layer so that, particularly in high amperage cells, a close approach of anode to cathode is not possible. If more were known about the movement of the metal layer, voltage could be reduced. The magnetic fields can be measured in the cell but there is a lack of suitable instruments for accurately measuring metal flow and metal layer position. Development within the industry of larger cells with their greater economy in operation has become a question of coping with the magnetic fields and their effects.

Improved performance over the years has been largely a question of better control but there still remains very little which can be measured on an operating cell other than voltage, and automation has to be based almost entirely on this. There is a great need for better monitoring instruments for temperature, interelectrode distance, rapid determination of bath composition, alumina content, etc.

With a 35% energy efficiency there is a lot of waste heat and the problems of recovering or using this have never seriously been tackled.

There are problems still waiting solutions in the recovery of waste materials, particularly the spent cell linings with their contents of alumina, carbon and fluorides. Some partial recovery of the fluorides is at present achieved but the carbon and alumina are dumped. Recovery of the valuable constituents is becoming very pertinent, and dumping is becoming embarrassing.

Alternative methods

The previous section discussed problem areas in the development of the Hall cell towards higher efficiencies but there is a need to escape from the present design where the low current density cell is spread out horizontally (because of the metal pool cathode). Large cell rooms and high capital costs are required for any given annual output, and heat losses are high. If an inert, permanent or dimensionally stable anode could be found there would be a much greater freedom in cell design. Electrolysis might be carried out between vertical plates so that much more production could be packed into a given volume and energy losses saved, or bipolar electrodes could even be used and several cells in series packed into one container. This is the method by which voltage is saved in the aluminium chloride electrolysis cell at present under development by Alcoa. If the same design could be applied to alumina-cryolite electrolysis there would be many advantages and the step of converting the alumina to chloride in the Alcoa process would be avoided. There is therefore a continuing search for a permanent anode and new cell designs, even though the depolarising action of the carbon would be lost.

There could be greater energy savings if the electrolytic process could be replaced altogether by an electrothermal reduction. Thermal reduction processes are theoretically attractive but no practical process has yet been developed.

Conventional semi-fabrication

Approximate energy consumption at various stages in the smelting and fabrication cycles is indicated by the data in Table 4·4. Complications arise in considering the total energy content of aluminium semis because:

- much of world aluminium production uses hydroelectric power, but some is based on fossil fuels;
- except at primary smelters, recycled scrap generally predominates in furnace charges.

Where semis are made from primary aluminium of non-hydroelectric origin, nearly two-thirds of the total energy input, starting from bauxite, is consumed at the reduction stage and the remainder divided fairly equally between production of alumina and fabrication. Production of liquid aluminium from scrap requires less than one-twentieth of the energy expenditure when starting from bauxite.

Table 4·4
Approximate energy consumption during various stages of the aluminium fabrication cycle[1]

Production stage	Energy input (GJ/tonne)	(MWh(th)/tonne)
Smelting (alumina to hot metal)	212	59[2]
Smelter casting	4·3	1·2
Fabrication plants (remelt cast shops)	9	2·5
Secondary smelting (foundry ingot)	14	3·9
Rolling	24	6·6
Foil (from foilstock)	18	4·9
Extrusion	25	6·9
Castings	48	13·2

1 Based on UK data collated by Aluminium Federation for the years 1974 and 1975. Figures given relate to the year with the larger throughput in each case, but the UK smelters were not all running at full capacity.

2 Includes hydroelectric power converted to MWh(th) at a factor of unity and power from fossil fuel converted to MWh(th) at actual efficiency of industry-owned power stations or at 29·3% for power purchased from the grid.

Melting

Metal to be cast into units for working is mostly melted and held liquid in fuel-fired reverberatory furnaces. These may be of single or double hearth type, with melter capacity up to 80 tonnes (equivalent in volume to nearly 250 tonnes of steel). In castings foundries, melting and holding are often carried out in crucibles of 1-tonne capacity or less.

Fuel demand is governed in the first place by the efficiency of the furnace design, which is usually low because in the past the main aim has been good materials handling. A good furnace design may attain 25% efficiency or, with the rarely used recuperation, 50%. Few reach this level and many are below 15%. Thus, whilst the theoretical energy requirement to raise aluminium from room temperature to 725°C is 0·3 MWh(th)/tonne, the fuel energy actually consumed in doing so is commonly five or more times as great. Efficiencies markedly below even 15% are common because of poor choice, siting and adjustment of burners, poor maintenance of burners and furnace structures, and particularly, inadequate consideration of furnace utilisation in production scheduling. It is not always appreciated that a furnace is working at zero efficiency when molten metal is held at constant temperature. This is common in cast shop (wrought) and foundry operation and, particularly in the latter case, can lead to overall thermal efficiencies well below 5%.

There is clearly great scope for the application of existing basic understanding of the processes of combustion and heat-transfer to the improvement of design and operation of practical furnaces. There is also scope for better instrumentation and monitoring. Such knowledge is applied to large energy using equipment, like chemical reactors and steam generators, but has hardly touched the problems of small users, who may also be suffering most from inefficient plant utilisation. A useful approach to improvement in fuel efficiency is to introduce some form of recuperation to heat incoming combustion air or metal, and increases in the price of fuel have made this step economically more attractive than formerly. Nevertheless relatively simple steps could probably be taken to increase thermal efficiency in many cast shops without replacement of the basic hardware already installed.

In size, the large reverberatory furnace may be approaching the stage of diminishing returns, and it has been suggested that faster and more economical melting might one day be possible with very large coreless induction furnaces. Meanwhile metal-circulating pumps are being used in some cast shops to increase melting rates in reverberatory furnaces, particularly where the melting of scrap is involved. Use of oxygen-assisted combustion has been inconclusively studied over many years.

The true economic balance between fuel-fired and electric induction furnaces is still to be resolved, and likely to be dependent on the physical nature of the scrap in the metal charge because of the importance of melting losses in any overall energy balance. Thus the copper industry has developed a shaft furnace which is very successful with a standardised input of copper cathode, but attempts to develop a similar furnace for aluminium gave rise to oxidation problems with fine scrap and with magnesium-containing alloys. There is need for a wider analysis of heat transfer in the heating of irregularly shaped pieces of metal in high velocity gas streams and of the behaviour during melting of highly oxidisable alloys of long freezing range.

Liquid metal treatment

Until recent years, it was customary in wrought cast shops to degas and clean liquid aluminium in a holding furnace, necessitating chlorine treatment for up to 1 hour, or settlement of the melt for 15 to 20 hours in a top-heated electric resistance furnace where conditions were favourable. The advent of methods for in-line degassing and cleaning, capable of coping with any quality of incoming metal, has enabled the minimum holding time in the furnace to be reduced to zero, with consequent increase in metal throughput and reduction in energy requirement per tonne. Metal losses are reduced (see Table 4·5), and the energy and materials otherwise expended in fume treatment are saved.

Table 4·5
Melting losses for one melting/treatment cycle

Operation		Metal loss (%)
Smelter holding furnaces		1·00
Remelt cast shop		
(mixed ingot/scrap charge)	Al and Al-Mn	1·50
	Al-Mg-Si	2·00
	High Mg	3·50
Scalpings/sawings		2–10[1]
Lacquered can scrap		10–30[1]
Heavy chlorine fluxing in holding furnace		0·40[2]
Metal loss in producing re-roll via conventional route compared with strip casting		0·70[3]

1 Depending on alloy and procedure.

2 Comprising 0·25% lost in the irrecoverable portion of the extra skim produced and 0·15% lost by chemical reaction with chlorine.

3 For basis see International Metals Review, 1976, **21,** p.111.

Casting

Casting is not in itself a particularly energy-intensive step but in direct chill (DC) casting it is clearly desirable, wherever practicable, to empty the holding furnace in a single drop, and this can involve pouring a number of blocks or billets simultaneously. The development of hot top, level pour casting systems has now greatly facilitated the multi-strand pouring of small billets.

Recent Russian casting developments involve the use of electromagnetic forces to support liquid metal prior to chilling (usually referred to as electromagnetic or mouldless casting). The extra energy requirement is considerable but the process should give overall benefits by eliminating surface scalping operations prior to hot working and reducing edge cracking in rolling. There is undoubtedly scope for reducing the depths of scalping cut in other ways, for example by insulated rather than water-cooled moulds.

Heating solid metal

Convective heating is established for heating aluminium blocks and higher circulation speeds give improved heating rates. Low temperature heads must be used to avoid overheating thin material since it is difficult to measure and control actual metal temperature. A few continuous heat treatment plants for strip are in operation outside the UK. Induction heating is established for heating extrusion billets but not so far for strip; this is partly again a problem of the measurement and control of temperature. Non-contact techniques for the measurement of surface temperature still leave something to be desired after 40 years of radiation pyrometers. Significant increases in sensitivity appear possible in the future using versions which measure radiation in two or more colour bands.

Considerable improvements in thermal efficiency have been demonstrated in batch heat treatment furnaces where the furnace shape, air circulation system and charge disposition have been designed in accordance with aerodynamic principles.

In heating for working, a distinction needs to be made between simple pre-heating to the working temperature and longer soaking at this or some other temperature with the intention of effecting a metallurgical change in the cast structure. Here there are important interactions between composition, solidification rate, and soaking temperature on the one hand and, on the other, workability, recrystallisation behaviour and the possible need for an annealing step in the course of the cold rolling process. For common grade sheet alloys, compositions have not yet been optimised to minimise the energy input in the 'homogenisation'* and annealing stages. Equally, little attention has been paid to the possibilities of processing heat treatable alloys in such a way as to reduce, by one or two orders of magnitude, the time required to achieve full solution heat treatment.

Hot and cold working

Working processes demand less energy than heating and are much less amenable to improvement. Flow stresses vary with composition and metallurgical state, and there is some scope for reductions in energy requirement along these lines. In extrusion, the initial upset represents a large fraction of the total energy requirement, so that any metallurgical step which reduces the upset pressure (whilst permitting development of full tensile properties) could be helpful.

Somewhat larger savings could probably be achieved by reducing the energy consumed uselessly in the working operations themselves. Examples are the use of water-based lubricants in cold rolling (which would permit

*In the aluminium industry the term homogenisation is applied to high temperature soaking treatments carried out prior to hot working with the object of improving working behaviour. The accompanying microstructural effects comprise dissolution and diffusion of some constituents and precipitation of others from supersaturated solution.

considerably greater reductions per pass), effective lubricants in extrusion (which would permit extrusion of longer billets), and perhaps the application of ultrasonic energy to tube-drawing dies.

Forming is done cold but flow stress could be reduced by hot forming (as with magnesium) and probably at little inconvenience. The overall energy position would need assessment, bearing in mind the cheaper tooling materials that could be used.

Recently a superplastically deformable aluminium sheet alloy has been developed. It is opening up new fields for the production by hot superplastic forming of complex formed shapes not previously possible. In some cases these are replacing fabricated structures composed of many parts.

Melting loss and yield of saleable product

Melting loss and yield of saleable product are important to the consideration of both energy and metal conservation. Because of the large amount of run-around in-house scrap associated with semi-fabricating operations (see Figure 4·1), metal charges in remelt cast shops comprise more scrap than primary metal.

Figure 4·1
Material flow diagram for production of rolled aluminium products

Source: Russell A S. Energy conservation in primary metals processing. Symposium on Controlling the Energy Delta (Demand-Supply). American Association (San Fransico). February 1974.

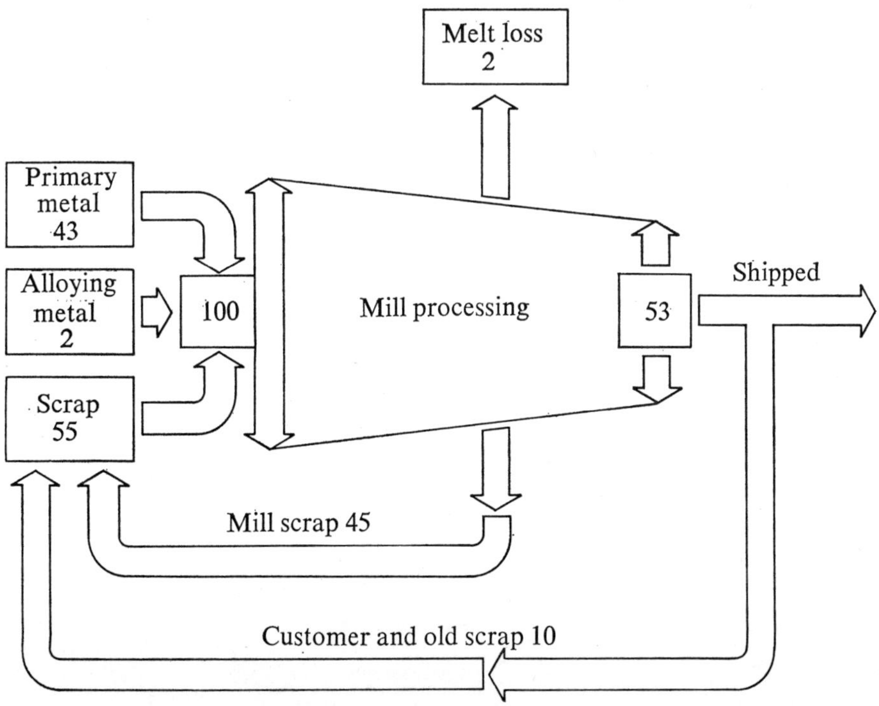

A distinction needs to be drawn between unit melting loss in a single remelting cycle and the overall melting loss, measurable over a long period in relation to the weight of primary metal fed into the system. If the average metal charge contains 2 parts scrap to 1 part primary metal and the unit melting loss is 1%, then the overall melting loss is 3%. The proportion of scrap used in the metal charge is essentially related to that generated and therefore decreases rapidly as the yield of saleable product per tonne of metal cast increases. For rolled circles and sand casting, yields will necessarily be low but purely on the grounds of minimising overall melting loss there is every incentive to maximise product yield. Some illustrative data on metal losses are given in Table 4·5; they are intended to be indicative only.

Metal losses represent an irrecoverable loss of the energy used in metal extraction, since there is no obvious contribution to the heat balance of the furnace.

Metal loss. Of the unit melting loss itself, part results from direct oxidation of metal in the atmosphere of the furnace, but much arises through formation of 'skim' (a floating, pasty mass of liquid metal and oxide films) from which the trapped metal cannot readily be released to unite with the rest of the metal bath. Skim is formed whenever metal is melted or liquid metal is alloyed or stirred. Its initial metal content when freshly formed in remelt furnaces is high but falls subsequently through oxidation to about 60 to 70% by the time the melt is skimmed prior to casting. Skim in the furnace reduces heat transfer by providing a partial barrier between metal and furnace roof. The layer is therefore heated disproportionately and may catch fire either in the furnace or on pulling, leaving a product with only 20% residual metal.

Improved methods of recovering metal from skim are needed. At present, the best approaches involve pulling the skim whilst still rich and either disintegrating it hot in liquid flux or cooling it rapidly to stop oxidation and then ball-milling, sieving and remelting the recovered metallics.

Attention should also be directed to preventing skim forming in ordinary reverberatory furnaces; skim formation can be excessive if materials like swarf, foil or lacquered scrap make up the charge. One solution with such materials is to use a furnace with a charging forewell on which a layer of liquid salt flux is maintained, and to charge the scrap into the furnace through the flux layer. In this way skim formation can be suppressed, and a salt slag is obtained containing about 10% of metal. However, although metal losses are considerably reduced, thermal losses from the open forewell must be significant. Without flux, skim formation from fine scrap can be reduced by immersing the scrap rapidly in molten metal, using induction stirring or a liquid metal pump.

The conditions likely to influence the formation and growth of skim do not seem to have received much study. Relevant factors are:

alloy
size and shape of scrap melted
thickness and degree of hydration of surface oxide films
presence of organic films or paper coatings
method and rate of melting
turbulence in the liquid metal bath
temperatures and times of holding
presence of magnesium in the bath
levels of the trace elements sodium, lithium and beryllium.

Product yield. Improvement of product yield offers a wide variety of studies —from details of process technology and materials behaviour to systems of process operation and scheduling and including techniques for in-process control of quality. Most factories are continually looking for improvements in product yield, involving optimisation of proceessing conditions to suit changes in input materials and product type. To achieve major improvements a bolder approach is indicated and investigation of alternatives or new process routes should be stimulated. For example, the extension of horizontal DC casting to products other than those for which it is at present used is worth examination.

Newer approaches to semi-fabricating

Bypassing the hot rolling line

Highly relevant to the problem of reducing energy consumption in working operations is the ability to cast strip directly, either thin strip (7 mm) for direct casting or thick strip (25 mm) for subsequent hot rolling and coiling. Both these processes are subject to some constraints in alloy composition and casting speed, but these are diminishing as a result of recent developments. At present, the existence of many hot rolling lines—comprising breakdown mill and multi-stand hot mill—largely precludes the use of strip casting for other than minor new or incremental capacity, but the proportion of rolled products made from strip cast metal appears likely to increase considerably in the future.

In addition to the reduction in the amount of hot working permitted by strip casting, there is also the elimination of the homogenisation/pre-heat step. Published invesigations agree that the total energy requirement to produce re-roll strip from liquid metal by the strip cast route is only two-thirds of that with the conventional route.

Strip casting is only one, but hitherto the most successful, way of bypassing the hot rolling line. Many thousands of tonnes of sheet have been made commercially from roll-compacted pellets. Spray casting on to a substrate has been intensively investigated. Fast horizontal DC casting of thin strip has been demonstrated on a production scale and could prove a way of overcoming some of the present limitations of thick and thin strip casting from moving moulds.

Bypassing the holding furnace at smelters

The advent of in-line degassing and cleaning has led to a questioning of the function of the holding furnace at smelters producing simple alloys. If metal were brought from the reduction cells in say 10-tonne travel pots, it would not appear a difficult matter to adjust its composition and temperature and to pour each in turn into a common launder leading directly through an in-line metal treatment unit to the casting machine. Not only would the fuel for the holding furnace be saved, there would also be a metal saving (and therefore an energy saving) corresponding to the metal otherwise burnt in the holding furnace or lost in the irrecoverable fraction of the furnace skimmings.

An integrated approach to the production of aluminium semis

Starting from a green-field site in the 1980s, and with energy proportionately more expensive than at present, how could the production of aluminium semis best be arranged so as to minimise energy consumption? What research and development would be desirable now in order to make such a scheme possible? One answer for the 1980s might incorporate the following features.

- Produce primary metal by chloride electrolysis, by the permanent anode technique or by direct reduction.
- Feed reduction cell metal through in-line metal treatment units to the casting machines.
- Eliminate holding furnaces or separate holding hearths in remelt cast shops.
- Use strip casting for production of almost all rolled products other than thick plate.
- Use Sendzimir or planetary type cold mill to produce large coils of wide thin strip from cast strip (probably in a single pass). The cast strip might be given a single warm pass before coiling for some products.
- Use horizontal DC casting for extrusion log.
- Use DC cast block and billet free from surface segregation for those products which cannot be made satisfactorily from cast strip or from horizontally cast billet: the block would have to be rolled on a conventional hot mill elsewhere. (Electromagnetic casting is one possible way of producing block and billet free from surface segregation.)
- Optimise the composition of common grade alloys (Al and Al–Mn types) to suit strip casting particularly to avoid the need for homogenising anneals before working.
- Arrange the composition and processing of heat treatable alloys to avoid the need (or greatly reduce the time required) for separate solution heat treatment.
- Use reverberatory furnaces with metal circulation for remelting scrap.

Equip them with coverable forewells for remelting swarf and direct flux-processing of hot skim.

Reclamation and secondary refining

Scrap recycled through secondary smelters is used almost entirely for foundry ingots because of their greater tolerance to impurities. Although magnesium can readily be removed by chlorine, and zinc less readily, so far no practicable refining technique has been developed to reduce the levels of iron and silicon (universally picked up in remelting) and other alloying elements, such as copper and manganese, to a level that would permit full re-use of the output in wrought alloys.

By present methods, the magnesium is removed by converting it to magnesium chloride and the energy input in producing this metal is wasted. Three-layer electrolysis would enable the magnesium to be removed from scrap and subsequently re-used. This type of process has been used to produce super purity aluminium for many years, but most of the applications for this special product can now be met with cheaper, less pure metal obtained by a freeze-thaw approach or even by selection from normal production.

Deliberate collection of easily identifiable wrought materials, such as beverage cans, is practicable with a good deal of unpaid effort on the part of the public, but the scope for this effort is limited.

Research effort is perhaps best devoted to scrap sorting and possible new approaches to chemical purification. An 'aluminium magnet' has already been developed for scrap sorting and, to improve chemical purification, attention might be given to the possibilities opened up by rheocasting (in which the metal is vigorously sheared when partly solid) leading perhaps to easier separation of impurity-laden eutectic liquid.

Downstream activities

Shaped castings

The high energy input needed for shaped castings production (see Table 4·4) is associated with the holding of liquid metal at temperature for long times in thermally inefficient furnaces and, except for high pressure die casting, with the generation of much in-house scrap. These two factors have already been discussed in connection with semi-fabrication.

Other opportunities doubtless exist for reducing energy input by, for example, moving to core mixtures and moulding processes which avoid the need for core baking operations and by using fully collapsible cores permitting re-use of core sand. Major energy reductions appear unlikely without improvements to thermal efficiency and the ratio of cast to fettled weight.

Foil

Foil rolling is carried out cold from coiled sheet (foilstock) using few passes on fast mills and with high product recovery. Consequently the opportunities for energy saving appear minimal.

Fabrication
Forming operations on aluminium are carried out cold with small energy involvement.

Wire drawing
Rod for wire drawing is usually hot rolled from either DC cast bar or from thinner bar cast continuously on a Properzi-type wheel. In the latter case the emerging rod is looped directly into a multi-stand mill and the pre-heating of cast stock thus avoided.

Recent developments such as the Conform process, in which coiled bar is forced through an orifice, and the Outokumpu process, in which bar is cast upwards from a melt, may well hold promise of future energy savings.

Copper
Production and consumption data for the UK are given in Table 4·6.

Table 4·6
Approximate production and consumption of copper in the UK[1]

	Copper	Quantity (tonne × 10³)
Primary refined	Production[2]	70
	Imported	350
Secondary	To refined copper	70
	Remelted	210
Semi-manufacture	Wrought as copper	420
	Wrought as alloy	270
	Foundry	70

1 Based on Metallgesellschaft data for 1974.
2 Produced in the UK from imported blister copper.

Smelting
There is no primary copper smelting carried out in the UK, but a considerable amount of copper scrap is refined to the standard of high purity primary.

For about 200 years, most copper has been produced from sulphide ores by controlled oxidation. The mineral is dispersed within rock and is liberated by crushing to a fine powder which can then be made to attach itself selectively to air bubbles by suitable additives and separated as a concentrate. Oxidation of the concentrate in large reverberatory furnaces or bottom-blown converters produces a liquid sulphide matte and then metallic copper. The energy is mainly supplied by the oxidation of the sulphur content of the mineral. In recent years, considerable effort has gone into the development of rapid and continuous processes to improve energy efficiency and plant utilisation, and particularly to reduce the volume of combustion gases so that the sulphur dioxide can be more effectively removed.

The resulting blister copper is further fire refined and then cast into slabs which become the anodes in electrorefining cells. The cells comprise a very compact assembly of alternate anode blocks and cathode sheets in acidic copper sulphate solution. Copper dissolved from the anode is slowly deposited as high purity metal on the cathode in a month-long cycle. Some impurities, like nickel, accumulate in solution; others, particularly precious metals, separate as slimes and are a valuable by-product. Much of the cathode copper is melted and cast into wirebars before shipping to industrial centres in other parts of the world.

Extraction is now increasingly from oxidised ores and requires a totally different approach. Concentration of the mineral is usually not possible, and the whole of the crushed rock must be treated with acid or ammonia to dissolve the copper values for subsequent chemical processing in aqueous solution. Increasingly, the copper is complexed with an organic reagent and selectively transferred to an organic solvent from which it can be redissolved in acid at a concentration suitable for separation by electrolysis, although this time using an insoluble anode and requiring eight times the energy needed for simple electrorefining of copper anodes. Any precious metal values must be extracted separately. These hydrometallurgical processes are in slow but active development and will become increasingly important. In the long term they may influence UK practices in recycling copper base materials. They may also lend themselves to the production of copper powder as a basis for new routes to some semi-manufactures.

Secondary refining

For re-use as wire, copper scrap is refined in the same way as primary blister copper. It is melted in large reverberatory furnaces (up to 200-tonne capacity) and impurities such as lead, tin and zinc are slowly oxidised out into slags. The traditional 24-hour refining cycle is unnecessarily prolonged; recent work at the BNF Metals Technology Centre has shown that impurities can be removed in minutes rather than hours by treatment with specific slag compositions of controlled oxidation potential. This technique offers the prospect of continuously refining a stream of liquid metal being fed to a continuous casting machine, for example.

Grossly impure scrap, no longer suitable for re-use as a melt addition, may be treated in a blast furnace whose energy requirements are supplied partly by coke but quite largely by the iron content of the scrap itself. Impurities are oxidised and slagged off and the crude copper produced is passed on to the fire refining furnace.

As in primary extraction, fire-refined copper from scrap is cast into anodes and further purified by electrorefining.

Conventional semi-fabrication

Melting and heating

In melting, a clear distinction must be made between copper and its alloys. Brasses for wrought products have long been melted in low frequency induction furnaces, giving controlled conditions with low zinc loss. Only in the last 10 years or so have these furnaces commonly been of more than 1-tonne capacity. Their melting efficiency is around 60% of the electrical input but less than 20% on the basis of the original fossil fuel input.

For moderate outputs, particularly in foundries, both copper and alloys may be melted in fuel fired furnaces—small reverberatories, Sklenars and crucibles—usually of very low efficiency but good operating convenience. Serious attempts are now being made to improve practices particularly by introducing more instrumentation.

Melting losses are often high but since, unlike aluminium, drosses can be fairly readily reduced back to metal, they are often not taken very seriously. Fumes, particularly of zinc oxide and carbon (from oil in swarf), must now be effectively collected, presenting many practical problems.

Preheating for hot working is still largely carried out in direct flame furnaces. These operate on counterflow principles which improve thermal efficiency, but many have inefficient geometry (designed for materials handling convenience alone) and antiquated burner systems.

Annealing after cold working is a common operation, often followed by pickling to remove surface oxidation. Pickling is increasingly costly because of requirements in effluent disposal but considerable attention is being paid to minimising the need for its use. Copper wire has long been annealed in coil in vacuo but there are problems from the welding of adjacent turns and much is now heated electrically in single strand, although at very low thermal efficiency (below 3%). Strip is increasingly annealed in strand by convective transfer from hot gases. Brasses may suffer surface deterioration from oxidation or zinc volatilisation unless dwell at temperature is very brief; in achieving this there are serious problems of temperature measurement and control.

Working

Copper and its alloys are usually dealt with in much smaller block sizes than aluminium but continuous processing is now well established in the production of wire. Alloy rod is made by extrusion, but the whole production of shaped sections is much less developed than with aluminium, because of the greater technical difficulties and the more limited range of applications. Strip is hot and cold rolled from traditional blocks (up to 2-tonne weight) but some alloys in small production are made by continuous casting and rolling at low speeds.

Energy demands are not high but product yields of 60 to 80% (apart from wire) offer scope for energy reduction by reducing in-house scrap.

Semi-fabrication by newer methods

Wire

During the past ten years or so almost all new installations for wire rod production have been based on continuous melting, casting and rolling to wire rod size (9-mm diameter).

The Asarco shaft furnace for continuous melting is virtually standard for the larger installations with capacities of 50 000 to 150 000 tonnes/year. The furnace takes a standard charge of cathode plates in packs and heat is applied to the bottom of a vertical shaft from specially designed burners injecting hot neutral gases through the descending metal. Melting is confined to a small zone near the hearth and can be started and stopped quickly to match the demand for liquid metal. High thermal efficiencies can be achieved. A holding furnace is necessary for storage and to superheat the liquid to the temperature required for casting. Further development of this type of furnace to cope with a range of alloy types and charge distribution would help many other industries.

Casting is carried out on several types of machine. Grooved wheel and belt or twin belt represent moving mould types, the Outokumpu machine casts vertically upwards through static dies, and the American 'GE Dip' process freezes liquid copper on to a seed wire passing upwards through a sealing die.

No hard data have been published but these continuous processes undoubtedly use less energy than the traditional hot rolling of a wire bar, with much of the benefit coming from the melting stage. They also give higher product yield.

Rod and section

Continuous extrusion of rod and section has long been sought. In a new process now under investigation (Conform), bar is forced through a die by friction in a grooved wheel, and a continuous output is conceivable from a continuous cast input or by welding together individual pieces of bar in the die. Current demonstrations on small aluminium sections point the way to the use of other materials and larger sizes.

Strip

A continuous process is probably most likely to give both lower energy demand and improved product yield.

Some alloy strip is already made by casting a rectangular section in a static graphite mould and rolling directly. A novel rolling mill giving 95% reduction in one pass has been used in one such installation.

For 40 years, thin copper strip has been made by electrodeposition on a revolving stainless steel drum. Although the production rate is low, the method is competitive with strip rolled down from conventional blocks. Engineering developments could make the process more widely applicable, particularly if based on scrap material.

Copper powders can be continuously rolled into strip and might be made available from scrap sources. Conventional powder rolling is restricted to moderate throughputs for technical reasons but other techniques, such as laying down a slurry on a steel substrate, seem capable of further development.

Recycling

Copper and its alloys are usually easily recognised in scrap by colour and mode of occurrence and have long been sufficiently valuable to encourage separation and return. The more obvious articles (such as cable, tubing, containers, radiators, electric motors, etc) are stripped from vehicles, appliances and buildings at the end of their service life and returned through established channels. Unlike aluminium, much general scrap is returned for remelting in a processing route which does not pass through a secondary smelter. For example, leaded machining brass is made almost entirely from returned scrap, including much machining swarf from product manufacturers. A steady degradation occurs during the cycles of re-use; copper is fed into ductile brasses, scrap from these into leaded brasses, then into casting alloys, and finally grossly impure scrap is refined back to high purity copper.

Much of the success of this operation depends on the recognition of particular alloys by the articles in which they are used. In the past, much of the recognition has been visual by trained sorters but more mechanised methods based on combinations of quantifiable physical properties must be found.

When recycled copper materials have been refined, the main aim has been the recovery of the copper content and impurity elements have largely passed into drosses and slags. There is now considerable interest in elaborating the refining techniques to recover these secondary elements but at the moment there is still no great economic return in achieving this.

Zinc

Approximate production and consumption data for zinc in the UK are given in Table 4·7.

Table 4·7
Approximate annual production and consumption data for zinc in the UK

	Zinc	Quantity (tonne × 10^3)
Primary	Production	80
	Imported	190
Secondary		40
Uses	Brass	90
	Galvanising	90
	Diecasting	70
	Rolled	20

Smelting

UK production of zinc is from the single Imperial Smelting blast furnace at Avonmouth, working with mixed zinc and lead sulphide concentrates. The sulphur is largely burnt off in a sinter strand and the blast furnace proper is charged with the mixed oxides, coke and a siliceous flux. Molten lead collects in the hearth and zinc vapour passes out with combustion gases at the top. A major problem is to prevent re-oxidation of the zinc by the gases during cooling and there has been considerable study of the thermodynamics and kinetics of the reactors to prevent this. In the furnace at Avonmouth, zinc vapour is rapidly condensed by a shower of liquid lead droplets, producing zinc liquid containing about 1·5% of lead in solution. This was once a common article of commerce but much of the production must now be freed from lead by distillation, preferably under vacuum.

Most of the world's zinc is, however, produced by electrodeposition from a solution of pure zinc sulphate, prepared from the roasted sulphide concentrate by an elaborate sequence of chemical manipulations. The metal is of high purity (99·99%), suitable for critical applications such as the preparation of pressure diecasting alloys but unnecessarily pure for many others.

The electrolytic route has a high energy demand (19 MWh(th)/tonne) whereas the blast furnace needs only 5 MWh(th)/tonne of zinc and produces in addition up to 0·5 tonne of lead. In both cases (unlike copper) the energy of the sulphur content of the mineral input is largely lost, apart from that usefully recovered by heat exchange from the flue gases. The sulphur itself is recovered as sulphuric acid which makes a useful contribution to process economics.

Melting and casting

Substantial quantities of zinc are melted for galvanising steel (see Table 4·7). Most of the energy demand in galvanising comes from heating the very much larger weight of steel to operating temperature and, for the continuous galvanising of strip and wire, in soaking at temperature for surface conditioning in furnaces of considerable length. Much of the zinc used forms a sludge of zinc-iron compounds, known as hard spelter but this is passed on for recycling.

The melting of zinc alloy for pressure diecasting heats much less metal but tends to be carried out in small furnaces at low thermal efficiency. Greater importance is attached to efficient metal handling than to efficient use of energy. As in most shape casting, up to half the metal is external to the casting proper and, although the excess metal is easily re-used, the energy demand of the product must include the recycled metal. The true product yield is often not known because process scrap is usually fed into the holding furnace without adequate record.

Flat-rolled zinc is used in this country mainly for making battery cans but in Europe it is used far more widely for roofing. It is now made relatively efficiently by a continuous casting route.

Reclamation

The uses of zinc tend to be destructive, like galvanising and battery cans, and they provide no positive basis for recycling. However, diecastings made from high purity metal can, with care, be recycled but they are usually degraded to some less critical use. Much general scrap and hard spelter from galvanising tanks is turned into saleable zinc oxide and zinc dust by distillation, thus saving primary metal.

Zinc oxide dust is a widespread by-product of melting processes, from brasses to remelting galvanised steel scrap, and the recovery of metallic zinc from the oxide has long been considered. Its physical properties and the way in which it arises in small amounts at widely scattered locations have so far kept such a process uneconomic.

Lead

Approximate production and consumption of lead in the UK market is shown in Table 4·8.

Table 4·8
Approximate annual production and consumption of lead in the UK

	Lead	Quantity (tonne $\times 10^3$)
Imports	Ore	45[1]
	Bullion	100
Recycled		200
Use	Batteries	100

[1] Lead content of ores.

Smelting

True smelting of lead in the UK is mainly confined to the zinc-lead blast furnace but a great deal of imported crude lead is refined here. The processes for the profitable separation of precious metals (mainly silver) by partition into zinc, and the removal of impurities such as copper, tin, antimony and arsenic by successive treatments with sulphur, air, caustic soda, etc, is carried out in large open kettles. They demand energy because of prolonged operating cycles rather than through any intrinsic energy need. Hence, present studies aim to increase reaction rates and achieve shorter cycle times.

Scrap lead is readily accepted in refining process routes.

Processing

Both melting and working have low energy demands and process scrap is easily returned for remelting.

Reclamation

Lead in everyday use is easily recognised and passes through well-defined channels for recycling. The return of metal and oxide in secondary cells is organised on a large scale and the lead content is recoverable as metal by melting with a reducing agent. The process is basically simple but there are many practical handling problems because of the associated plastic cases and free sulphuric acid. For some time into the future, modified refining practices will have to be used as maintenance free batteries are introduced in which the usual antimony addition is replaced by calcium and tin.

Nickel

Some statistics on world and UK use are given in Table 4·9.

Table 4·9
Annual consumption of nickel in the UK and the Western World, 1974

	Nickel	(%) (tonne × 10^3)	Usage Quantity
Primary consumption	Western World	715	
	UK	34	
Pattern of use	Nickel plating		16
	Nickel-base alloys		10
	Copper-base alloys		3
	Iron-base alloys		60
	Stainless steel		41
	Constructional steel		10
	Castings		9
	Other		11

Smelting

Early production of nickel was by treatment of oxidised ores but these were later replaced to a large extent by sulphide ores, whose treatment paralleled that developing for copper sulphides. Nickel sulphide ores can be concentrated and smelted to sulphide mattes and then to metal. Electrorefining can follow, if desired, giving profitable concentrations of precious metals in addition to the nickel.

In the past 20 years or so there has been increasing development of processes for oxidised ores. In the long term most nickel will come from lateritic ores where somewhat different treatments are required for the mixed oxides occurring near the surface and the deeper silicates. A variety of process routes is used, some converting the ore to sulphide for pyrometallurgical treatment and some dissolving the mineral in acid or ammonia with individual stages designed to suit particular deposits. The crude metal may again be refined by electrolysis or by making use of the unique property of nickel in forming a gaseous carbonyl. The latter is the basis of the only operation within the UK.

Heating and hot working

Nickel-base materials have melting points similar to those of steel and are melted in arc furnaces, induction furnaces or, for better control of specialised alloys, in vacuum-arc or plasma-arc furnaces. All these have relatively low thermal efficiencies.

The sizes of melt and cast products are small compared with steel and aluminium, but not compared with many copper-base processes, and blocks for working are cast in static moulds. Some attempts have been made at continuous casting but the billet sizes needed and the very low thermal conductivities of some alloys are unfavourable.

Many mixed alloys are designed for high temperature service and are in consequence particularly resistant to deformation in hot working. To help overcome this, extrusion billets are glass lubricated but still deform the dies, whilst forging has to be carried out rapidly with many reheats. Metallurgical factors override concern for simple energy efficiency.

Recycling

Nickel is a relatively expensive metal and reclamation is important.

In the production of nickel-base materials, careful segregation of types allows in-house recycling of most process scrap. The smaller quantities of mixed materials, such as machining swarf, are remelted, ingotted and sold as a source of nickel for less critical applications. Nickel in pickle liquors is precipitated and the dried sludge returned to the primary smelter. Nickel alloys used in well identified applications, such as aerospace components, can be recovered and returned.

Nickel as an electroplated coating usually makes up such a small proportion of an article that no attempt is made to recover it. However, it may become an embarrassment to the steel remelter. Drag-out from plating solutions must be precipitated and was often dumped, but there is now some attempt to recover nickel and recycle water through the use of reverse osmosis or ion exchange. With both these techniques, improvements are desirable.

Warren Spring Laboratory studied for some years the recovery of alloying elements in nickel-base materials but no commercial use has so far developed. There could be scope for the development of solvent extraction techniques in the general removal of nickel from solutions and these might in the longer term find application in primary extraction.

Nickel associated with copper, as alloy or coating, may ultimately appear in the electrolyte of any secondary refining process in the form of a sulphate. Some hundreds of tonnes are recovered annually in this form.

Within the steel industry, stainless scrap can be recognised and re-used, and structural steels can be recycled as such when in pieces of adequate size. In the long term, however, nickel is steadily degraded in continuous recycling until it becomes an undesirable impurity (like copper) in mild steels. The pattern and extent of recycling are not deducible from published statistics.

Titanium

The annual level of world consumption of titanium is around 50 000 tonnes including 3 000 tonnes in the UK.

Details of applications are not published, but their early dependence on aerospace is declining and commercial use of unalloyed metal in chemical equipment and heat exchangers will predominate in future. At present there is a marked growth in the production of castings, some up to about 100 kg in weight.

Extraction

The standard form of titanium metal is as sponge—a granular product made by the reduction of titanium tetrachloride with a reactive metal such as magnesium. USA, Russia and Japan dominate world markets. In the UK, titanium granules are made by Imperial Chemical Industries by dripping the tetrachloride into molten sodium and removing the sodium chloride with water. Their production is in 2-tonne batches at present but as demand grows it might be desirable to develop a continuous operation. The Electricity Council has studied the continuous decomposition of titanium tetrachloride in a plasma arc.

Semi-fabricating

To produce billet, the granules are mixed with alloying powders and some particulate scrap and are compacted into blocks which are then plasma-welded into long electrodes for vacuum melting. The liquid pool is continuously solidified in a copper mould cooled by water or sodium-potassium eutectic. This energy is lost as low grade heat. The solid billet is scalped and may be vacuum remelted for metallurgical improvement or to give a larger diameter billet.

Forging to bar or rolling to flats normally requires several stages with intermediate heatings at 900 to 1 100°C. In one installation, heating is at present carried out at low efficiency, but remedial measures are being considered.

Energy demand in subsequent cold deformation is relatively small. In particular, the energy demanded in tube production will fall as tube welded from strip takes over from the solid drawn product.

Overall yield is about 60% of the granular input and recirculating scrap is around 20%.

Recycling

In-house scrap, largely machining swarf, is broken down and incorporated into the electrodes. Samples are checked by X-ray fluorescence analysis. Ferrous inclusions are separated magnetically and most other contaminants are readily removed by passing through a fluidised bed of suitable effective density.

Scrap generated elsewhere is not at present taken back for recycling, partly because the likely pattern of contamination is unknown. Some at least is used elsewhere for making master alloys.

Magnesium

Some statistics on world and UK use are given in Table 4·10.

Energy consumption in the magnesium sector is dominated by two factors, the high input required to produce the primary metal and the wastage due to metal losses in melting operations.

Table 4·10
Annual production and consumption of magnesium in the UK and the Western World, 1976[1]

	Quantity (tonne × 10^3)	Uses (%)
Metal supply (Western World)		
Primary	207	
Secondary	21	
Market distribution (Western World)		
Aluminium alloys		48
Chemical/metallurgical		27
Structural		25
Despatches (UK)		
Ingot	4·4	
Wrought forms	1·95	
Powder	0·5	
Castings	0·45	
Other products	0·1	

[1] Data from publications of the International Magnesium Association and the Magnesium Industry Council.

Smelting

Since primary production is not carried out in the UK, the energy input requirement is not directly of concern here. However, it may be noted that developments are in hand with both the electrolytic and the silicothermal extraction routes which could bring the energy input per tonne of primary magnesium down to the level of aluminium. Present total energy input is 105 MWh(th)/tonne.

Heating and hot working

Magnesium-base alloys have melting points similar to those of aluminium alloys but are not melted in conventional reverberatory furnaces owing to their high chemical reactivity with steam and refractories. Melting is carried out in iron pots of up to 2-tonne capacity in conjunction with salt fluxes containing magnesium chloride. Thermal efficiencies are comparable with those applicable to crucible furnaces in the aluminium foundry industry.

Metal losses arise partly from oxidation and partly from metal entrapment in the flux residue remaining at the bottom of the pot after the metal has been poured. With coarse scrap in the charge, metal losses per cycle are 2 to 3% for Mg-Al type alloys and 5 to 7% for magnesium-zirconium alloys. With swarf scrap, recoveries of 70 to 90% are usual, the value depending on the state of the swarf. Current work on the fluxless melting of magnesium, in which an atmosphere of sulphur hexafluoride is used, could lead to reductions in metal losses in some circumstances.

Energy requirements for the pre-heating and heat treatment of magnesium alloys are generally comparable on a volume basis with those for aluminium and thermal efficiencies of the furnaces used are similar.

Recycling

There are few technical problems in the remelting of scrap and, subject to chemical limitations, all kinds of scrap can be remelted to give alloys of similar quality and properties to those made from virgin ingot. Magnesium alloy scrap is recovered from aircraft and road vehicles and swarf and sawings are returned from machining shops. Only fine grindings cannot be recovered at present and have to be burnt under controlled conditions.

Future place of magnesium

Although magnesium is at present highly priced in relation to aluminium, it is recognised that economies from current developments could lead to much increased use in transport applications. The energy benefit of substituting light metals for steel, measured over the life of the vehicle, is some 80% greater for magnesium than for aluminium. The methods of saving energy in the extraction and processing of magnesium have been considered recently at an international conference.*

General improvements in metal processing operations

Energy

As an aid to further thought about energy saving, a list is given below of the features of the processing of metals which appear most conducive to energy economy. Whilst some of these features have already been adopted in the processing of certain metals, others may offer further opportunities for eliminating energy-intensive steps, the need for which may not always have been adequately challenged. Changes, of course, must be environmentally acceptable.

- Where the primary metal is generated hot, cast directly into shapes for subsequent working (i.e., without metal treatments, settling periods, etc in holding furnaces).
- Make castings of geometry to minimise subsequent working.
- Hot work cast shapes without intermediate cooling.
- Control composition and solidification rate to minimise the need for subsequent isothermal soaking treatments to effect structural changes for improving workability or developing strength.
- Minimise melting losses by
  - reducing metal loss per melting cycle,
  - reducing amount of in-house scrap by increasing product yield (metal sold/metal cast),
  - minimising or eliminating surface machining operations prior to working.
- Optimise thermal efficiency of all melting and heat treatment furnaces by

*International Conference on Energy Conservation in Production and Utilisation of Magnesium, held at Massachusetts Institute of Technology, Boston, Mass., May 25–27, 1977; M C Flemings et al.

attention to design, fuel/air ratio, leaks, damper operation, materials scheduling, use of waste heat for preheating metal charge and incoming air, etc.
- Minimise use of holding furnaces operating at near zero efficiency.
- Recycle suitable metal from industrial scrap, municipal wastes, etc.

Materials

Several comments and ideas in the field of material conservation are generally applicable.

- More efficient application of materials in both design and manufacture of articles reduces metal demand per product unit.
- Improved product yield in manufacture has more effect on energy demand and plant utilisation than on the basic demand for metal.
- Improved recycling of old scrap is urgently in need of study, calling for
  - more collection of used articles,
  - better sorting to meet the needs of the recycling industries,
  - where possible, extension of refining back to primary metal standards,
  - maximum recovery from scrap in both remelting and refining.
- Sorting of recycled material involves the separation of metal content from residues, before or after pyrolysis, and the further separation into different metal fractions by exploiting differences in physical characteristics such as ferromagnetism, density, electrical conductivity or crude fusibility. The increasing use of mechanical assemblies and composites (such as plated articles, galvanised steel and copper-coated aluminium wire) to give optimum service performance adds to the burden of the scrap recycler.

University research

The following notes list some industrial needs, problems and processes which may provide the basis of suitable topics for university research.

Aluminium

Reduction
- Investigation of mechanisms leading to current inefficiency in the Hall cell.
- Anode reactions, for example, basic investigation of over-voltage.
- Low resistance electrolytes.
- Measurement and prediction of the movement of metal layers in high currents and high magnetic fields.
- Improved monitoring equipment for measuring alumina content, temperature, and inter-electrode distance—all under working cell conditions.
- Recovering waste heat.
- Radically new process concepts, for example, permanent non-consumable anodes.

- Development of a practical and economic electrothermal reduction process.

Casting and working

- Improved method for melting and holding aluminium compared with the existing non-recuperative reverberatory furnace—preferably incorporating continuous charge feeding with counter flow of combustion gases.
- Study of the basics of containing molten aluminium by an electromagnetic field (mouldless casting).
- Alloy development for minimum energy input in homogenisation and annealing.
- Development/optimisation of low-alloyed, non-heat treatable composition for casting processes involving very quick solidification, for example, continuous casting of thin strip (6 mm) through moving moulds.
- DC casting of strip in heat treatable alloys and other compositions of wider solidification range through stationary moulds.
- Study of factors influencing the oxidation of aluminium and the growth of skim in melting, alloying and transfer operations.
- Improved methods of recovering metal from furnace skim.

Secondary smelting

- Methods for upgrading scrap to primary quality metal (particularly silicon and iron removal).
- Improved methods of sorting scrap.

Copper

Secondary refining

- More rapid reaction to shorten cycle times.
- Attention to useful recovery of other metals present as alloy or impurity.

Melting and other heating operations

- General improvement in use of furnaces by
  - basic design,
  - better operation,
  - use of waste heat.

Newer routes

- Development of continuous melting furnaces of wider application, with analysis of basic factors in heat transfer and in melting behaviour.
- Continuous casting of simple shapes.
- Production of forms other than cathode plate in electrorefining.
- Production of strip from powders.

Recycling

- Better methods of identification and separation of alloy types.

Zinc

Melting and casting

- Better design and use of furnaces in melting.
- Better throughput and product yield in the use of pressure die casting machines, mainly by more instrumentation and reactive control and more thorough analysis of the design of castings.

Recycling

- Increased recovery of zinc from the fumes arising from zinc-coated steel scrap and brass melting (containing zinc oxide), which at present are not economically treatable.

Lead

Refining

- Reduction of energy demand by more rapid reactions and more efficient furnaces.

Nickel

Recycling

- Improvement in recycling of plating solutions.
- Better methods of identification and sorting of solid articles in various nickel-base materials.
- Improvement of solvent extraction techniques for secondary or primary nickel in solution.

Titanium

- Making castings direct from the melting process.
- Casting a rolling slab.
- Casting small diameter bar in long lengths.
- Continuous casting of shapes for working.

(In all these there are basic problems in linking the casting stage to the continuous vacuum melting and in the last, at least, in matching melt rate and freezing rate.)

- More efficient methods of pre-heating for working (may be by induction heating).

Magnesium

- Method of cleaning liquid metal without use of salt fluxes.

General

In all melting and heating processes, particular attention needs to be given to the recovery of waste heat, preferably by pre-heating incoming materials. The recovery of low-grade heat contained in gases and cooling waters is a general problem.

5 Rubber and plastics

Chapter contents

Introduction

Plastics and rubber products form an increasing proportion of the material used in the UK and have been shown (National Economic Development Organisation, 1974) to be in the higher grouping of energy intensity, being 9th, 22nd and 25th in the 85 industries listed.

Size of the industries

The level of UK production in 1976 was of the order of 2·5 million tonnes of plastics products and 1·1 million tonnes of rubber products. These totals are composed of polymers, either synthetic or natural together with various additives, fillers, reinforcing media, plasticisers and chemicals. The total for plastic polymers was about 2·04 million tonnes (European Plastics News, 1977) and for rubbers (including latices) about 0·58 million tonnes (Rubber Statistical Bulletin, 1977). It is difficult to give precise figures for the total tonnages of plastics and rubber products because a number of the products are composites of polymeric materials with other classes of materials such as rubber/metal bonded components, coated textile products and plastic coated materials.

Some indication of the total size of the industries in financial terms can be obtained from the Business Statistics Office (Business Monitor, 1977). Data for 1976 derived from these sources is presented in Table 5·1. Two totals are given: the gross figure includes the value of the basic polymers and the net figure is the value of the saleable products to the consumer market.

Table 5·1
Value of UK plastics and rubber industries in 1976

Minimum list heading	Coverage		1976 total (£m)
Plastics			
276–1	Synthetic resins and plastics materials	(polymers)	985
		(products)	358
492	Linoleum, plastics floor-covering, leathercloth etc.		244
496	Plastics products not elsewhere specified		1 313
Gross total	Plastics products		2 900
Net total	Plastics finished and semi finished		1 915
Rubber			
—	Natural rubber		88
276–2	Synthetic rubber		160
491–1	Rubber—tyres and tubes		549
491–2	Rubber—other rubber goods		798
Gross total	Rubber products		1 595
Net total	Finished rubber goods		1 347

Again, this list does not give the overall value as certain items are listed elsewhere. Examples of these are: textile fibres from plastics materials, formulated synthetic resin adhesives, plastic toys etc, plastic mattresses, plastic cable insulation, rubber footwear, rubber weatherproof outerwear, rubber mattresses, rubber adhesives and rubber thread. Total sales for these additional products are of the order of £200 million.

Distribution of products

Table 5·2 illustrates the very wide distribution of the use of plastics and rubber materials across industry. Major identifiable areas of use for plastics and rubbers are based on the 1976 UK sales figures (Business Monitor, 1977).

Table 5·2: UK sales of plastics and rubber products in 1976

Plastics	1976 total (£m)	Rubbers	1976 total (£m)
Packaging containers	334	New tyres and tubes	504
Film, sheet, rod	196	Hose and tubing	90
Building products	185	Belting	76
Flexible and rigid foam	95	Mouldings	64
Leathercloth and sheet	88	Sheet products	48
Household fittings	68	Retreaded tyres	45
Electrical components		Adhesive film and tape	40
(excluding cable)	68	Metal bonded products	40
Domestic holloware	65	Rubber compounds	38
Floorcovering	57	Rubber seals	24
Laminated sheet	42	Cellular products	23
Vehicle components	35	Other uses	203
Other uses	235		

These figures illustrate the dominance of packaging uses for plastics (particularly as much of the film is used for wrappings) and of the tyre industry for rubber.

Nature of the industries

The industries can conveniently be considered in three sections—the manufacture of the basic polymeric material (usually classified as part of the industry but fundamentally a chemical industry), the forming processes for plastics, and the forming processes for rubbers.

Polymer manufactures

A part of this first section of the industry is concerned with natural products: the natural rubber imported from various tropical countries, naturally occurring resins and cellulosic materials. These are all produced by biological processes occurring in vegetation which are converting, with the use of solar energy, basic chemicals from the air or the ground into polymeric materials based on carbon, hydrogen and in some cases oxygen. Further energy inputs are required to convert the natural materials into a suitable form for use.

The rest of the products are based on chemical synthesis of the polymer starting from a suitable raw material, normally oil. Here again further energy inputs are required to bring about the polymerisation and to convert the product into a powder, crumb or sheet suitable for use by the conversion sectors of the industries into saleable products.

Plastics processors

In the plastics conversion industries two main types of material are involved: polymers on their own or polymers compounded with fillers, colouring, plasticisers and reinforcement. The use of material with fillers requires the input of additional energy to bring about the blending of the ingredients. The use of just the polymeric material avoids this energy input as such additives as may be required (for example, stabilisers) will normally be incorporated in the final stages of the polymer production and, in any case, represent only a small proportion of the total material.

The feed to the forming section is then processed into a shaped product by a variety of means: extrusion, calendering, injection moulding, compression moulding, blow moulding, film blowing, vacuum forming, dough moulding and slush moulding. In all these cases, energy input is required to carry out two functions. First it raises the material to a temperature at which flow can be achieved and second it transports the material into the mould. This is usually done under pressure to ensure adequate filling and consolidation of the product. Following the forming operation the heat present in the polymer has to be removed to allow setting of the shaped product and to permit easy removal from the mould. This also involves energy input but does allow the use of heat recovery.

In injection moulding processes a certain proportion of the material is wasted, as the mould requires runners to convey the material into the mould cavity and the content of such channels is surplus to the final product. In compression and blow moulding an excess is present as sprue. This is essential to ensure complete filling of the mould. In other forming processes, waste results from trimming or cutting of the product. This excess can easily be recycled if the polymer is thermoplastic, usually requiring only the energy needed to granulate the scrap into a suitable form for re-use. The recycling of thermosetting materials is more difficult because chemical reactions have been involved resulting in a material which is different from the original feed to the process.

Rubber processors

Processing in the rubber industry can be conveniently divided into three sections: the compounding and mixing stage, the forming stage and the vulcanisation stage. These are normally separate although some continuous processes have been developed and are being used.

The compounding and mixing stage comprises the bringing together of the various ingredients involved. Some energy input may be required in preliminary processing, for example, in warming the bales of raw natural or synthetic rubber and in cutting the bales into more conveniently sized pieces. The mixing operation frequently involves the mastication of the polymer by shearing the polymer in mixing equipment using considerable quantities of energy. This is followed by the addition of other components, again using energy to shear the mixture and make it homogeneous. Heat is generated in this process and the equipment is frequently cooled. Any waste produced at this stage due to incorrect mixing, incorrect formulation or other causes can usually be recycled with minimal difficulty.

The compounded material is then passed to the forming stage. The typical processes employed in this stage are: extrusion, calendering, moulding (either by compression, transfer or injection), spreading, and dip coating. In most of these, energy inputs are required to drive machinery and heat inputs are sometimes needed to lower the viscosity of the material for easy processing.

The final stage is that of vulcanisation in which heat input is required to initiate and maintain the chemical reactions leading to the cross-linking of the compound. In some cases, for example moulding operations, this is combined with the shaping operation. The energy requirements here are usually considerable and are discussed later. An advantage of rubber moulding is that the strength of the hot product and its elasticity permit the removal of the product from the mould without cooling. Waste from this process, additional to faulty products, is in the form of flash and excess of vulcanised rubber from runners, mould joining lines, etc. which must be trimmed from the product. This waste is not suitable for immediate recycling as it has been chemically changed.

Energy involvement in the industries

Energy is of considerable importance to the plastics and rubber industries, and is described here in the same sections used in the introduction with the addition of sections dealing with energy in waste material and environmental energy needs.

Energy requirements for polymer production

Plastics

A variety of data has been published on the energy requirements of a number of plastics polymers and other materials. The figures are summarised in Tables 5·3, 5·4 and 5·5.

Comparison of the data in Tables 5·3, 5·4 and 5·5 shows very clearly the different values for the energy requirement of the same feedstock material reported from different sources. These differences may be due to variations in the technology used, the scale of operations, or the degree of energy conservation exercised at the plants where the measurements were made. They may also be due to differences in the boundary conditions chosen.

The range of values emphasises that statements of energy requirements of materials or products need to be made with great care. This particularly applies when making comparisons based on single values for energy requirements covering a broad subject, such as manufacture of the feedstocks for containers.

Table 5·3
Energy requirement for feedstock material (Freshwater, 1975)

Material	Density (g/cm³)	Energy requirements Feed stock (GJ/tonne)	Conversion (GJ/tonne)	Total (GJ/tonne)	By volume (MJ/cm³)
Aluminium	2·7	–	246	246	0·66
Steel billet	7·8	–	44	44	0·34
Tinplate	7·8	–	55	55	0·43
Copper billet	8·9	–	53	53	0·46
Bottle glass	2·4	–	20	20	0·05
Paper & board	0·8	–	62	62	0·05
Cellulose film	1·45	–	194	194	0·29
PVC	1·38	24	62	86	0·12
Polystyrene	1·07	57	83	140	0·15
LDPE[1]	0·92	49	50	99	0·09
HDPE[2]	0·96	50	53	103	0·10
Polypropylene	0·90	52	61	113	0·10

[1] Low density polyethylene
[2] High density polyethylene

Table 5·4
Energy requirement of feedstock material (Data Base, 1974)

Energy source	Energy requirements Raw Steel (GJ/tonne)	Glass[1] (GJ/tonne)	PVC (GJ/tonne)	LDPE (GJ/tonne)	HDPE (GJ/tonne)
Coal	18·1	7·5	8·7	–	–
Oil	1·5	1·6	18·7	25·7	28·9
Gas	3·0	10·3	53·6	73·1	75·3
Electricity[2]	1·8	3·1	22·6	19·7	8·7
Total	24·4	22·5	103·6	118·5	112·9
By product fuel credit	2·1	1·4	7·3	9·9	9·9
Net total	22·3	21·1	96·3	108·6	103·0

[1] The basic material, glass, is produced in the physical form of the container.
[2] Based on an energy requirement of 10·84 MJ/kWh delivered.

Table 5·5
Energy requirements for feedstock material (Plastics and Rubber Research Institute TNO, 1974)

Energy source	HDPE (GJ/tonne)	PVC (GJ/tonne)	PE coated paper: Paper (GJ/tonne)	PE coated paper: PE (GJ/tonne)	Glass (GJ/tonne)
Fuel	26·6	17·0	5·8	2·0	11·7
Steam	0·4	15·0	35·4	1·0	–
Electricity	1·0	25·4	3·2	0·2	0·7
Total	28·0	57·4	47·6		12·4
Energy equivalent of raw material	47·0	23·3	23·2		–
Total	75·0	80·7	70·8		12·4

Some data (Freshwater, 1975), indicating the differences between theoretical and practical energy requirements for the production of monomers and polymers in the UK, are summarised in Table 5·6.

Table 5·6
Annual energy requirement for some monomers and polymers

Chemical	Energy requirements: Theoretical ($GJ \times 10^6$/yr)	Energy requirements: Actual ($GJ \times 10^6$/yr)
Ethylene	+5·44	+21·2
Vinylchloride	−0·63	+ 9·0
Butadiene	+5·06	+ 9·67
Styrene	+0·042	+10·55
Polyethylene	−1·84	+17·9
PVC	−0·67	+14·85

However, data of this type are very difficult to ascertain with an accuracy better than a crude order of magnitude.

Rubbers

Little data have been published concerning energy requirements for natural rubber production but some figures (Malaysian Rubber Producers Research Association, 1977) are shown in Table 5·7.

Table 5·7
Energy requirements in natural rubber production

Process	Energy (GJ/tonne)
Crepe preparation	0·25 – 0·38
Crumb drying	4·2 – 4·3
Transport, Malaysia-UK	1·5

For the synthetic rubbers, data are available (Data Base, 1974) for styrene-butadiene rubber (SBR) which represents the greatest use of polymer in the industry (see Table 5·8).

Table 5·8
Energy used in synthetic rubber production

Energy source	Energy requirement (GJ/tonne)
Energy for electricity	13·67
Refined oil products	66·57
Natural gas	75·14
Total	155·38
By product fuel credit	16·01
Net total	139·37

The same reference (Data Base, 1974) gives a figure for the production of reclaimed rubber of 24·5 GJ/tonne.

Energy requirements for conversion to plastics products

Present requirements

Much of the data available for this section refer to specific studies made on particular products. The first few refer to the production of plastics container with a capacity of 0·25 to 2·0 litres and compare the energy requirement of various alternative materials. The energy required to produce a variety of 0·34-litre containers is shown in Table 5·9 (Bunt, 1974).

The table immediately illustrates that the energy consumed by glass despite its comparatively greater density, is conserved by the number of times that the container can be re-used.

Table 5·9
Energy requirement for container production

Container	Energy requirement (MJ/container)
Returnable glass bottle (1 trip)	6·1
35 g plastic bottle	4·8
Non-returnable glass	4·2
Tinplate/aluminium can	3·6
25 g plastic bottle	3·5
Returnable glass bottle (3 trips)	2·9
All tinplate cans	2·8
Returnable glass bottle (8 trips)	1·5
Returnable glass bottle (19 trips)	1·1

Plastics often consume less process energy than the material they replace even if the hydrocarbon raw materials are counted as energy and the heat value of the product is not considered. Their comparatively larger energy requirement is the result of counting the feedstock material resources used as energy rather than as a raw material. In plastics production, 1·41 tonnes of crude oil make 1·38 tonnes of naphtha which gives 1·05 tonnes of ethylene from which 1 tonne of polyethylene is produced (Imperial Chemical Industries Ltd, 1974 and Spencer-Palmer, 1974).

Table 5·10 (Bunt, 1974) gives a useful breakdown of the energy consumed during container production.

Table 5·10
Energy requirement for container production

Production stage	Energy requirement (MJ/kg)		
	Glass	Plastics	Tinplate
Conversion to containers	4·4	17·6	4·4
Heating and lighting	1·8	7·0	1·8
Transport of containers	0·9	0·9	0·9
Raw material production	13·2	101·0	44·0
Total	20·3	126·5	51·1

Further data from the literature have been collected by TNO, Netherlands and are summarised in Table 5·11.

Table 5·11
Energy requirement for plastic container production

Capacity (litres)	Weight (kg)	Energy (kWh/kg)	Energy (MJ/kg[1])	Material
3·78	0·08	0·925	11·1	HDPE
1·90	0·134	1·28	15·4	PVC
2·00	0·08	0·96	11·5	PVC
1·00	–	1·03	12·4	PE/PVC
0·45	0·019	1·05	12·6	PVC

[1] Based on the conversion factor of 12 MJ/kWh.

From the same source, data for the production of 1-litre containers from feedstock material are shown in Table 5·12.

Table 5·12
Energy requirement for container production

Energy source	Energy requirement (MJ/kg) HDPE	PVC	PE coated paper	Glass
Fuel	–	–	–	0·4
Electricity	12·4	12·4	4·72	3·53
Total	12·4	12·4	4·72	3·93

The data from TNO sources together with information on the energy content of the oil equivalent of the raw materials and on the potential recovery of energy by incineration are given in Table 5·13 in a summarised form.

Table 5·13
Total energy consumption for 1-litre containers

	Milk bottle HDPE	Plastic bottle PVC	PE-coated cardboard milk container Paper	 PE	One-way glass bottle Non-carbonated glass	 Carbonated glass
Weight of container (kg)	0·030	0·05	0·03		0·034	0·515
Weight of material for 1 kg of containers (kg)	1·01	1·07	0·802	0·221	1·00	1·00
Energy consumption for manufacture of material (Total 1)						
Fuel (MJ/kg)	26·6	17·0	5·81	2·02	11·7	11·7
Steam (MJ/kg)	0·40	15·0	35·4	0·97	–	–
Electricity[1] (MJ/kg)	1·14	28·1	3·56	0·25	0·77	0·77
Energy consumption for manufacture of 1 kg of containers (Total 2)						
Fuel (MJ/kg)	–	–	–	–	0·4	0·4
Electricity[1] (MJ/kg)	12·4	12·4	3·6	1·12	3·53	3·53
Total 1 + Total 2 (MJ/kg)	40·5	72·5	52·7		16·4	16·4
Total 1 + Total 2 (MJ/container)	1·22	3·63	1·58		5·58	8·47
Energy equivalent of raw materials used						
Raw material energy (MJ/container)	1·43	1·17	0·44	0·31	–	–
Energy recoverable by incineration						
Energy recovered (MJ/container)	1·36	0·88	0·44	0·26	–	–
Total energy consumption (MJ/container)	1·28	3·91	1·63		5·58	8·47

[1] Based on a conversion factor of 12 MJ/kWh.

Further summated data, for tinplate and tinplate/aluminium containers of 1-litre capacity from TNO are shown in Table 5·14 on a comparable basis to Table 5·13.

Table 5·14
Energy requirement for 1-litre metal containers

Container	Weight (kg)	Total energy (MJ/container)
Tinplate	0·13	9·11
Tinplate with 30% returned scrap	0·13	6·91
Aluminium-ended tinplate	0·12	10·47
Aluminium-ended tinplate with 30% returned scrap	0·12	8·05

As part of an energy survey made for the EEC, practical measurements of the conversion processing energy for containers were made (Schaefer and Watson, 1975). The outcome of these, giving ranges, are as shown in Table 5·15.

Table 5·15
Measurements of energy for container plastic processing

Polymer and container sizes (litre)		Energy usage MJ/1 000 units	MJ/kg polymer	MJ/m³ capacity	MJ/unit bulk density (kg/litre)
PVC	0·35–1·00	92–300	2·7–7·5	120–675	1·7– 5·9
HDPE	0·60–1·35	168–478	4·6–9·4	195–840	3·4–11·0
LDPE	0·60	180	5·2	320	3·0

The inclusion of the various units is necessary so that the particular end-use of the data can be accommodated.

Further data (ICI, 1974) refer to a variety of products but combine the energy requirement for the feedstock with that for the conversion to give a total energy figure (see Table 5·16).

Table 5·16
UK energy requirements for fabricated products

Product		Energy requirement for feedstock and conversion to product ($GJ \times 10^3$)
1 million m² packaging film		
	Polypropylene	4·84
	Cellulose	6·82
1 million fertiliser sacks		
	Polyethylene	20·70
	Paper	30·80
100 km of 1-in diameter service pipe		
	Polyethylene	2·51
	Copper	2·90
	Galvanised steel	10·20
1 million 1-litre containers		
	PVC	4·27
	Glass	10·12
100 km of 4-in drainage pipes with fittings		
	PVC	15·85
	Asbestos cement	17·60
	Cast iron	86·70
	Clay	22·00
	Pitch fibre	19·35

The lower energy requirements for the plastics materials are largely due to their lower density.

Reductions for the future

Increasing interest in processing energy reduction is evidenced by a number of publications in recent years. In one article (Murakami, 1975), the thermal efficiency for extrusion is claimed to be about 30% and for injection moulding 15%. Modified processing procedures are suggested to improve this utilisation. Studies of energy use in extruding (Yokoyama, 1975) indicate reduction in energy by improved machine design (see Table 5·17).

Table 5·17
Energy reduction by redesign of processing machines

Material	Energy (kWh/tonne) Conventional design	New design
HDPE	204–412	242
LDPE	163–232	215
Rigid PVC	163–232	144
Soft PVC	123–162	60
Polystyrene	136–204	107
Polypropylene (PP)	163–326	126
Acrylonitide butadene styrene (ABS)	182–325	152
Filled HDPE	290–380	250
Filled PP	270–350	220

Work is in progress at the National Engineering Laboratory (Gardener, 1977) to study the energy effectiveness of the hydraulic drive systems frequently featured in the industries. A study of an injection moulding machine gave the energy balance shown in Table 5·18, suggesting that installation of centralised hydraulics would give an energy saving of 67%.

Table 5·18
Energy balance of injection moulding machines

Injection moulding machine	Power consumption (kW)
Average,	800
less heater demand	200
Electric motor,	600
less standing losses at 67%	400
Demand to work machines	200

As a precursor to improvement the Rubber and Plastics Research Association is engaged, by the Department of Industry Engineering Materials Requirements Board, on contracts concerned with the energy use pattern in the major plastics and rubber forming processes. Preliminary indications from these investigations suggest that the actual requirement of energy for heating and forming is not more than 10% of actual usage. Some losses are thought to be associated with thermal loss to the environment and others in the drive systems as described above (Gardner, 1977).

The energy required for pelletisation has been theoretically analysed (Johnson, 1976) to enable optimisation of energy use to be achieved.

Energy requirements for conversion to rubber products

Less data are available in this sector than for the plastics products. Data collected from two sources (Malaysian Rubber Producers Research Association, 1977 and tyre manufacturers, 1977) breakdown the energy needs as shown in Table 5·19.

Table 5·19
Energy requirement to convert to rubber products (tyres)

Process/component	Energy (GJ/tonne)
Milling	0·65
Extrusion	0·16
Vulcanisation	0·5–3·2
Tyre ingredients	143
Tyre processing	26

Other data (Cheater et al, 1976) for tyre manufacture, which represents about half of the total rubber production, refer to the energy pattern associated with a typical car tyre weighing about 7 kg and show the approximate energy values:

energy of ingredients:	1 000 MJ
energy of manufacture:	185 MJ
total for new tyre:	1 185 MJ

Some data (Dannenberg, 1975) on the major filler use in rubber, carbon black, indicate a requirement of 6·8 litre of crude oil to produce 3·3 kg of product or approximately 75 MJ/kg. Energy use efficiency of 32 to 36% of theoretical requirements is claimed, dependent on the grade of black produced.

The possibility of reducing energy needs in rubber processing has been investigated in processing procedure and in the use of processing aids. A paper reviewing the use of single and two stage mixing (Roger and Roebuck, 1975) indicates reduction of power consumption from 12·5 to 8·9 GJ/tonne for styrene butadene rubber/butadene rubber (SBR/BR) mixes and from 14·5 to 10·2 GJ/tonne for natural rubber/butadene rubber (NR/BR) mixes by using single stage mixing. An industrial data sheet (American Cyanamid Co., 1976) claims reductions of peak load in mixing cycles from 27·5kW to 18·5kW by the use of chemical plasticisers in natural rubber mixing.

Energy involvement in waste materials and products

Plastics and rubbers are both high in energy if regarded as a fuel and also have large quantities of energy involved in their production. The importance, in energy terms, of recycling of materials in preference to using them as a fuel to recover their thermal energy content is shown by the figures in Table 5·20.

Table 5·20
Comparison of energy input for a polymer or product and its fuel value

Material/product	Energy content Materials (GJ/tonne)	Energy content Production of product (GJ/tonne)	Approximate calorific value (GJ/tonne)
Polyethylene	75–109	11–12·4	39
PVC	81–102	11·5–15·4	25
SBR polymer	135	–	37
Rubber compound	90–130	10–25	30
Tyre (worn)	143	26·5	22

Clearly, if the costs of recovery are not excessive, dumping a product is wasteful of its potential energy value because its value as a fuel is comparable with that of other normal fuels.

The recycling of materials can be considered, first, as an in-house operation and, second, as a return of used and rejected material.

Recovery of in-house waste

In-house recovery is fairly straightforward in the case of thermoplastic products. Assuming that arrangements have been made for separation by colours and material, the energy involvement is limited to converting the material into a suitable form (usually by granulation). Some energy measurements have been made at RAPRA on polypropylene as a representative material. The requirements of energy were from 0·1 to 0·6 GJ/tonne for one type of machine and from 0·6 to 1·2 GJ/tonne for another, the level being dependent on the rate of feed as the no-load energy for the equipment is high (60% at light loading, dropping to about 33% at maximum loading). Normal commercial practice is to have continuous recycling of waste from the process back into the production line.

Recycling of thermosetting plastics and compounded rubbers is more complicated as these materials have been chemically changed in the production line. A minor exception occurs in the intermediate processing of rubber where unvulcanised compound arising as trimmings can be immediately recycled in many cases or faulty batches of material can be adjusted and remixed. The re-use of other materials requires considerable equipment to regenerate the material in a useful form. The in-house process normally used is the grinding of the product to form a powder or crumb which contributes little more than a filler in recycling situations. Energy requirements have been determined for rubber compounds (Cheater et al, 1976): about 0·7 GJ/tonne to produce a powdered material.

Return of used and rejected material

The economic return of used products is dependent on the separation of the used material into appropriate categories if the arisings are in the form of rubbish. In some cases, separation is simple. Used tyres are frequently collected by the replacement tyre suppliers and are not contaminated with other waste. In the case of plastics and other rubber goods, however, the arisings are normally mixed. Experiments in certain countries have attempted to separate the rubbish at the domestic source, the dust bin (Bureaux Internationaux Techniques, 1976). Typical composition of such arisings is shown in Table 5·21.

Table 5·21
Composition of domestic refuse

Type of waste	Amount (% by weight)
Dust and cinders	12
Vegetable matter	17
Paper	43
Metals	9
Textiles	3
Glass	9
Plastics	5
Unclassified	2

Having obtained a reasonably homogeneous waste, the recycling can be attempted by using the grinding techniques used in in-house recycling operations.

An alternative is to reclaim the material and an energy level of about 7 GJ/tonne is needed for this operation. A Reverzer process has been developed in Japan for plastics (Wolk et al, 1974) but no details of energy requirements have been found. This process is however claimed to be capable of producing a material suitable for manufacturing products without further use of raw materials.

Another method of re-using worn or damaged products is refurbishing by the application of new material, as in tyre retreading. The report (Cheater et al, 1976) on waste tyre re-use indicates energy levels for this process of up to 58 GJ/tonne, including the preparation of the used tyre and the material and work required to apply the retread.

Failing the possibility of recycling the waste products, recovery of the chemicals contained in the material by pyrolysis, for example, is desirable. One study of pyrolysis (Cheater et al, 1976) attempts to identify the energy involvement for processing waste rubber tyres. Other centres (for example, the Warren Springs Laboratory) have been studying the pyrolysis of municipal refuse. An American plant (Wolk et al, 1974) produced 370 tonnes of carbon, 124m^3 of naphtha, 420m^3 of gas oil and 128×10^9 m^3 of sulphur free gas per day from 1 000 tonnes of rubber scrap at an energy cost of $6{\cdot}35 \times 10^9$ GJ/day of fuel, 2 640 kWh of electricity and energy for cooling water and boiler feed. This shows an energy deficit and can only be justified by other considerations. A similar approach has been tried for plastics but no indications of energy involvement have been traced.

Incineration

Incineration has been studied for rubber, and suitable furnaces are available. The potential for energy recovery in UK, based on the 22 GJ/tonne quoted in Table 5·20 and an estimate of waste tyres unsuitable for retreading or other recycling in 1974 (Cheater et al, 1976), is about 178 000 tonnes or a potential energy source of about 4×10^6 GJ per annum. The same source gives results for an installation where weekly savings of about 35 m^3 of fuel oil are being realised.

The burning of plastics waste can similarly yield significant thermal energy but some plastics, for example PVC, require special furnaces to cope with the corrosive gases evolved. Arisings in this case are more likely to be widely dispersed and, as such, would most probably be considered as part of the incineration of general refuse. European data (Bureaux Internationaux Techniques, 1976) suggest arisings of about 250 kg/year of refuse per capita and, taking the percentage analysis quoted in Table 5·21, arisings of plastics waste in the UK are likely to be of the order of 0·75 million tonnes or about 25 million GJ/annum. This would indicate that about one third of the total UK plastic production ends up in the dustbin.

Energy uses in the factory

In the preceding discussions, little breakdown of the function of the energy other than, for example, into raw materials and processing has been ascertained. The boundary within which the energy is measured is not usually defined but it is assumed that the boundary is the factory or the process and therefore includes environmental needs for energy.

Some indication of the breakdown of energy uses is being established as a result of Department of Industry Energy Audits and Energy Thrift investigations. The plastics and rubber industries have been visited, with one day in the factory, as part of the Industrial Energy Thrift Scheme (IETS). The coverage for plastics in MLH 276 (see Table 5·1) has been to 72 plants out of a potential of 178; visits to MLH 496 (other plastics products) have still to be made; and in MLH 491 and MLH 492 (rubber and coated textiles), visits have been made to 97 plants out of a potential of 416. Analyses of the data collected have not yet been completed but preliminary analysis has given some indications of energy breakdown and levels. This preliminary data is summarised in Table 5·22.

Table 5·22
Energy use breakdown—plastics and rubber

Energy	Chemical manufacture of polymers	Plastics processing	Rubber processing
Total input (% of total)			
Electricity based	15	35	25
Oil/gas/coal based	85	65	75
Electrical input (% of installed electrical load)			
Main drives	25	50	55
Compressors	8	10	10
Transfer	38	4	6
Process heating	1	18	15
Process cooling	20	4	2
Environmental	8	14	12

From evidence gained at the factory visits so far, the average annual usage of energy for the industries, together with indications of the proportion involved in environmental applications, is shown in Table 5·23.

Table 5·23
Energy levels breakdown

MLH description	No. of plants visited	Average energy per plant (GJ)	Energy for environmental control (% of total)
276 Plastics	70	385	
Resins + polymers	23	383	9
Moulding	8	25	38
Foam products	5	33	15
Film making	15	1 074	4
Compounding	10	69	28
Spreading	3	39	25
Laminating	6	138	23
491 Rubber	17	234	
(1) New tyres	6	598	19
Retread tyres	11	36	12
(2) Other rubber goods	61	73	22
492 Coated textiles	12	335	36

Preliminary indications suggest that some of the more important factors for energy conservation in the industry are (not in order of magnitude):

environmental control,
building insulation,
control of space heating,
recovery of waste heat from air circulation,

insulation and temperature control of domestic heat requirements,
control of lighting (for psychological reasons),
control of motor sizing and use,
control and maintenance of compressed air supplies,
control and maintenance of hydraulic systems,
insulation of heated process equipment,
recovery of heat from contaminated fluids from processes,
optimisation of boiler operations,
recovery of heat from cooling systems,
control of use of electrical demand and power factor.

Materials involvement in the industries

Plastics

Quantity used

The main component of plastics products is the polymer. This represents about 75% of the total weight of products manufactured. The amount of polymers used in the UK (European Plastics News, 1977) is shown in Table 5·24.

Table 5·24
Polymer consumption in the UK, 1976

Polymer	Consumption (tonnes × 10^3)
LDPE	417
HDPE	120
PP	185
PVC	422
PS	160
Expanded PS	19
ABS	36
Acrylics	23
Polyamide	20
Polyester film	14
Amino resins	136
Phenolic resins	60
Polyester resins	58
Epoxide resins	13
Polyurethanes	71
Others	287
Total	2 041

Indications are given in the same reference of the main markets of the larger use materials.

Precise data for the tonnage of other components are more difficult to establish but various sources (Business Monitor, 1977) suggest the following

approximate figures for 1976:
fillers 160 000 tonnes
reinforcement 80 000 tonnes
additives (including plasticisers) 220 000 tonnes.

In-house recycling
Conservation of materials can be considered from two aspects, the optimisation of the use of the material in the production stages and the recycling of material after use.

The optimisation of use of material is influenced by the choice of materials for the function (different materials require differing supplies of feedstock and energy) and by the effectiveness of the processing to minimise scrap. Recycling in house is not too difficult for thermoplastics but requires considerable effort for thermosetting materials. Increasing interest is being shown in the use of diluents. Data and predictions for the USA market (Modern Plastics International, 1975 and Tunbridge, 1975) are given in Table 5·25.

Table 5·25
Actual and predicted use of fillers in plastics in the USA

Material	Tonnage of filler (tonnes $\times 10^3$)		
	1975	1980	2000
Alumina	50	200	1 600
Asbestos	180	350	1 700
Carbonates	700	1 500	9 000
Cellulosic material	40	90	500
Glass	5	15	200
Silica	25	80	500
Silicates	6	15	100
Talcs	40	200	1 800
Miscellaneous	14	50	500
Totals	1 060	2 500	15 000

These figures are based on total production of plastics products of about 12 million tonnes in 1975, 18 million tonnes in 1980 and 60 million tonnes in 2000. Examples of the use of fillers are 40 to 80 parts (per 100 of polymer) of talc in polypropylene, 150 parts of calcium carbonate in polyethylene, and 70 parts of quartz microspheres in polyesters.

Some indication of the importance of additives used in the UK industry is given in price terms in Table 5·26 (British Plastics Federation, 1977). The values are compared with a total value of plastic material of £550 million in 1975. Additives represented 16·5% of that total.

Table 5·26
Value of plastics additives in the UK, 1975

Additive	Value (£ million)	(% of total)
Additive lubricants	2·677	2·9
Blowing agents	1·269	1·4
Catalysts and hardeners	2·213	2·4
Dyes and pigments	14·832	16·3
Fillers	16·880	18·6
Flame retardants	7·615	8·4
Plasticisers	36·640	38·1
Stabilisers	8·660	9·5
Others	2·100	2·4
Total	90·886	100·0

Increasing the proportion of such additives will benefit by reduction of the demand for the oil based polymers.

Some indication of the level of scrap material produced in the industry is given in a paper on the recycling of plastics materials in the USA (Bureaux Internationaux Techniques, 1976). The general level of in-house waste is about 10% and levels for various processes are estimated as shown in Table 5·27.

Table 5·27
Estimates of in-house scrap in the USA

Process	Amount of scrap (% of total product)
Pipe and profile extrusion	1–2
Sacks from tubular film	1–2
Bottle blowing	2–10
Flat film and sheets	5–20
Injection moulding	5–25
Vacuum forming, etc.	15–35

Lowering of these levels would lead to conservation of materials, particularly where direct recycling cannot be applied. However, repeated recycling leads to a deterioration in their physical properties.

Re-use of scrap

The second phase of material conservation covers the re-use of used scrap plastics products. As indicated earlier, the major difficulty lies in the collection and separation of the materials and some of the efforts made to overcome this are discussed there. There are particular problems associated with the use of recycled plastics (Bureaux Internationaux Techniques, 1976).

The problems are concerned with:

repeated recycling,
potential contamination of recovered materials,
limited compatibility of different plastics,
inconsistency of feedstock,
unpleasant odours due to decomposition of trace foodstuffs in used food containers.

On balance, the most likely route for reclamation from refuse, with the present state of technology, is not to extract material but to recover energy.

Rubbers

Quantity used

Rubber products are, with a few minor exceptions, unlike many plastics products in that the material used is not just a polymer but rather a complex mixture of polymer with other added ingredients. The base of the mixture is the polymer or a blend of polymers which may be natural products or oil-based synthetic products. The added ingredients may be particulate or fibrous fillers, extenders or processing aids (often oil-based products), chemical additives designed to improve the performance and ageing properties of the product and chemical additives needed to bring about the cross-linking of the mixture during the vulcanisation process which is required to give reasonable permanence to the product.

The magnitude of 1976 consumption of rubber is shown in Table 5·28 (Rubber Statistical Bulletin, 1977).

Table 5·28
UK consumption of natural and synthetic rubber

Material	Consumption (tonnes $\times 10^3$)
Natural rubber	168·3
Natural rubber latex	16·2
Synthetic rubber	327·4
Synthetic latex	49·6

The relative importance of the main synthetic rubbers used is not known accurately but an indication of UK levels is given by the relevant plant capacities shown in Table 5·29.

Table 5·29
Synthetic rubber production capacity in the UK, 1976

Rubber	Production capacity (tonnes × 10^3)
SBR	362
BR	40
EPDM	20
IIR	44
CR	23
NBR	21

The level of use of fillers, etc. is also not known precisely but various sources (e.g. Business Monitor, 1977) suggest the following approximate figures for 1976:

fillers 335 000 tonnes
additives (including plasticisers) 87 000 tonnes
reinforcement 90 000 tonnes.

Some indication (Wolff, 1977) of the types of fillers used is:

carbon black 57%
silica 5%
others 38%.

Precise UK figures for carbon black are available with a usage of 205 000 tonnes for all outlets and a total of about 190 000 tonnes for the rubber industry.

In-house recycling

The possibility of recycling rubber in house is very restricted, being limited to the return of unvulcanised products to the processing stages. As with plastics, this recycling must not be done too often as the part-processed rubbers contain the ingredients for vulcanisation which will take place at normal room temperature (although at a very slow rate) leading to an unacceptable material.

The avoidance of waste, trimmings, etc. is an important conservation measure. The best choice of material has been an accepted function of the rubber technologist for many years but the choice of the best for conservation of materials must now be considered as an additional responsibility.

Re-use of scrap

Once the material has been vulcanised, the problem of recycling becomes more complex. The situation has been analysed in depth for tyres (Cheater et al, 1976). Since tyres represent nearly half the production of rubber and are also the most easily collected used material they must present the most likely form of recycling feedstock. Recycling by crumbing, reclaiming or pyrolysis (Cheater et al, 1976) can also be applied to most forms of vulcanised rubber scrap.

The processes for tyres are retreading, use of scrap material for other functions, manufacture of crumb and reclaim, and pyrolysis for material recycling.

Retreading. Retreading is an important method for re-using worn tyres. A new tread is applied by a variety of techniques which provide about 75% of the depth of tread found on a new tyre. Bearing in mind that only 50% of car tyres and 65% of truck tyres are suitable for retreading the opportunity for growth of this recycling procedure is smaller than might be thought.

Retreading is currently undergoing world-wide technological changes, in particular in the adoption of prevulcanised tread techniques. Small companies have problems in keeping abreast of the latest technology. Methods exist for the control of casing quality, curing behaviour, the physical properties of tread rubber, final inspection and quality assurance but a need exists for a low cost instrument for the quality control of used casings for retreading.

Other methods for whole tyres. Some minor entrepreneurial activities make products from used tyres cut into shaped pieces, which take advantage of the plied fabric construction. They use specific parts of the tyre and discard the remainder.

Crumb and reclaim. Other than retreading, the major use for used tyres is through the manufacture of crumb and reclaim. Both processes require maximum separation of the rubber compound from the reinforcement, destroying the expensive tyre construction and producing low grade fibre and steel which have very few markets. Crumb and reclaim, although possessing particular and useful processing properties, are mainly used as compounding ingredients because they are cheaper than existing alternatives. In their present form it is not anticipated that major increases in use will arise, due to the reduced performance which results from higher percentage additions.

Increased use can be foreseen of a product that is capable of addition into tyres at higher levels. RAPRA has found that additions of up to 50% of the rubber in a tyre tread may be made if the crumb is reduced to around 300 mesh and all large particles are eliminated. If this is confirmed by testing, a large market can be seen for the product.

In the USA, commercial interests are promoting cryogenic methods for particle size reduction. The economics of the process rest largely on the cost of the refrigerant, generally liquid nitrogen. RAPRA is examining an ambient temperature process of its own invention. A techno-economic study to assess the most efficient way of obtaining crumb from a whole tyre would assist in clarifying the position.

Pyrolysis. A new approach for re-use is by pyrolysis. In various forms it is seen as a way of deriving more basic products. It is possible to control the process at laboratory scale to produce different proportions of three products—oil, gas and soot. The uses of these products are generally external to the rubber industry although possibilities have been cited for production of a type of processing oil and a type of carbon black.

No production plants exist as yet but there is a need to consider pyrolysis as a disposal process which is likely to be more environmentally acceptable than tipping or incineration. Much seems to depend now on the ability to avoid the use of any feedstock smaller in size than a quartered passenger tyre, due to the costs of granulation.

The order of magnitude of this potential recycling is indicated by the data shown in Tables 5·30 and 5·31 (Cheater et al, 1976).

Table 5·30
Disposal of used car tyres in the UK, 1974

Disposal method		Quantities	
	(units)	(tonnes × 10^3)	(% of total)
Total used tyres	23 467 588	159·580	100
Tyres in vehicle scrapyards	5 345 000	36·346	23
Tyres in reclaimers	5 294 120	36·000	22
Tyres exported	1 158 000	7·874	5
Tyres retreaded	3 737 588	25·416	16
Tyres rejected by retreaders	3 737 588	25·416	16
Remainder	4 195 292	28·528	18

Table 5·31
Disposal of used truck tyres in the UK, 1974

Disposal method		Quantities	
	(units)	(tonnes × 10^3)	(% of total)
Total used tyres	3 394 319	135·773	100
Tyres in vehicle scrapyards	632 000	25·280	19
Tyres to reclaimers	–	–	–
Tyres exported	261 812	10·472	7
Tyres retreaded	937 319	37·493	28
Tyres rejected by retreaders	504 652	20·186	15
Remainder	1 058 536	42·342	31

Tyres sent to vehicle scrapyards, tyres rejected by retreaders and the remainder item (totalling 178 000 tonnes) have no identifiable secondary recycling use and are presumably burnt or dumped. Currently, UK crumb and reclaim are only produced from fabric reinforced tyres although steel reinforced tyres are used for this purpose on the continent.

For used rubber products other than tyres, the only portion which is likely to be easily recycled is that arising in the rubber factory where collection and delivery to the recycling organisation can be arranged. Waste material from refuse collection, as with plastics and dependent in the same way on suitable separation techniques, appears to be suitable only for the recovery of heat value.

Recommendations for improved conservation

To set the background to this section it is worth considering the magnitudes of the opportunities for conservation. The largest single opportunity is recycling of the useful content of collected waste. This is because the plastics and rubber content of such arisings must represent the bulk of the plastics and rubber products which have completed their useful lives. The potential is, at least, the present total annual production of the materials or something of the order of 4 million tonnes. The largest quantities of plastics are in the form of packaging materials, followed by film and building products. Rubber is dominated by tyres, followed by hose, tubing and belting.

Conservation measures should first be directed at the materials and energy recovery from used plastics and rubber and then at the technology involved in the manufacture of the major products. The latter should consider all stages from the production of the raw materials used by the converters through to the final product.

It is important in considering the possible factors to relate them to the overall size of the industry. Materials will normally represent a greater proportion of turnover than will energy but these proportions are changing with continuing rises in energy prices.

Rubbish

The major problem to be resolved in the utilisation of waste must be the effective separation of the useful components from the mixed and variable feedstock. The cost of this in energy and financial terms must be significantly less than for the use of new materials (unless a shortage of materials allows a premium to be acceptable). In the case of plastics, problems may be that mixed polymers, for example PVC and PE, will be present in the separated plastic waste leading to compatibility problems. This problem may be less than anticipated, as the bulk of the arisings will be from plastic packaging, based almost entirely on polyolefins and PVC which can possibly be separated by flotation. The separation problem could be eased by careful selection and formulation of the materials with the objective of making separation from other waste more straightforward. Other factors which could aid recycling may not require technological developments but rather a change in the arrangements for waste collection—an administrative and political matter. A further factor which could justify investigation is the extent to which diversity in the recovered material could be tolerated whilst still permitting its re-use.

The large volume of rubber arisings, namely tyres, will not normally follow the route to the waste disposal organisation, as has been discussed earlier. The problem with this waste is to maximise the availability of the waste products for the established recycling procedures.

Raw materials

The manufacture of polymers, fillers and additives is mainly done by highly competitive industrial concerns normally using advanced technology. Energy and material conservation appear to have been an important consideration to them for some time but further energy conservation measures could be considered. IETS visits have shown a potential for waste heat recovery and recycling (usually low grade energy) and a need for the reassessment of quality control levels taking into account the energy cost of achieving a particular level. In the case of heat recovery the possibility of contamination is frequently a barrier to recycling.

Developments to reduce the proportion of polymer in the finished product by increasing the use of extenders have been described and could merit further investigation, with the possibility of both material and energy conservation. The latter could arise from lower thermal capacity of the mixture leading to energy economy in subsequent processing. Further studies of the design of products, including the use of composite products, could lead to further material conservation.

Continuing studies should be made to develop alternative materials which avoid some of the energy expenditure involved in alternately heating and cooling the material during production. Other possibilities arise if the physical form in which the polymer is supplied could be changed. It is possible that the polymer used in film manufacture could be produced in the form of film or sheet to eliminate further processing from granules or powder. This technique is already applied in the case of some cellulosic film production. Further studies could be justified in the production of large tonnages of oil-based carbon black to investigate alternative forms of input to the oil or gas normally employed. Waste products may be suitable.

Conversion

Much of current compounding is by batch processes but studies to perfect continuous operation could lead to energy and material conservation. Continuous operation could also lead to economy of energy and materials in other processes in the industry by avoiding waste in the start-up and shutting-down operations.

In both compounding and forming operations, additional studies of equipment design could lead to significant savings in energy by reduction of heat loss to the environment and in materials by the reduction of in-house scrap production. Material savings in this respect could also be anticipated from improved process control. Re-appraisal of the present technology could lead to a reduction in the number of stages in multi-stage operations between which cooling of material occurs. This loss of heat could be avoided in a number of instances. Studies have already been instigated into the energy efficiency of drives to processing equipment and in the recovery of heat associated with the drives. Such studies merit further investigation as considerable quantities of energy may be employed.

In forming processes, the supply of heat to lower the viscosity of polymers to ease their flow is a recurrent factor. Data have been presented which show that this can be a highly inefficient process in energy terms and investigations to improve the situation are in hand and should be continued. Energy inputs for melting polymers can represent a highly significant portion of the factory energy load. Studies to establish the current levels (such as the IETS scheme) and investigations to reduce losses from such inputs could achieve useful savings.

Finally, in the compilation of this chapter, there appeared to be a serious scarcity of background data concerning energy involvement and materials conservation in the industries. Much more data probably exist but these are likely to be confidential to particular companies. The collection of more data is desirable, covering all the major sectors within the industries, so that the authorities have available the necessary background information on which to base policy decisions. Some help in this direction can be expected as an outcome of the Department of Industry Energy Audit and Energy Thrift schemes.

University research

In considering the suitability of subjects for university research, the particular strengths of research and development potential in industry, industrial research associations and universities are important. The industry and its associated research and development organisations have particular and specialised knowledge of the science and technology of their own industry with a leaning towards the practical application. A weakness, particularly with smaller organisations, is the lack of time and resources to spend on research and development related to materials and energy conservation. In the past, this subject has had a lower priority than subjects more obviously linked with productivity and profitability. The universities have more generalised and diverse knowledge and have better access to other disciplines which may often be of relevance to the problem. In addition, the availability of time and resources may well be more easily achieved than in industry. Taking these points into consideration it would appear that the national interest will best be served by the continuation of active co-operation between all organisations to ensure the best use of all available assets.

Developments made, for example, in the rubber and plastics industries may have some relevance to other industries. In all this research it is important to back up the theoretical work with evidence from industrial situations, some of which could come from schemes such as the Energy Audit Scheme and the Industrial Energy Thrift Scheme. The following topics could be considered for university research.

Conservation of energy

- Theoretical energy requirements of major industrial processes. To extend over the production of polymers and major fillers from basic raw materials, mixing and compounding procedures, and major forming processes (for example, calendering, extrusion, compression moulding, transfer moulding, injection moulding, blow moulding, vacuum forming and vulcanisation techniques).

- Better techniques for the recovery of waste heat particularly when contaminated. To consider air extraction systems for vulcanisation equipment, reinforced plastics lay up, resin impregnation processes, solvent based spreading techniques, and drying processes. To consider condensate return systems for chemical reactors (where potential contamination by leakage is present), autoclave vulcanisation, and processes using live steam injection.
- Better techniques for the upgrading of waste heat. To use the large volumes of low grade heat lost in cooling and ventilation systems.
- Improved efficiency of energy use in providing drives. To study the significant energy losses associated with large drives required to meet intermittent high loads in processing machinery and with services such as compressed air.

Conservation of both materials and energy

- Theoretical effects of changes such as wear of machinery on the energy and material efficiency of major processes. To investigate the loss in effectiveness due to alterations in clearances in such equipment as internal mixers, extruders and injection moulding machines. Data are required to establish the time for machinery reconditioning or replacement for energy and material optimisation.
- A study of acceptable levels of diversity in recovered polymeric materials. To measure levels of dilution by, for example, paper and the extent to which mixtures of polymers can be tolerated. This is important to the levels of separation that are required of collected waste.
- Recovery of materials such as solvents, monomers, etc. in total loss extraction systems and the recovery and recycling of heat contained in such systems.

References

American Cyanamid Co.
Increasing rubber production with reduced energy costs
Rubber Chemical Lines **25** (4), October 1976

British Plastics Federation
Background report for Additive Suppliers Group
Press Release, September 1977

Bunt B P
Energy cost of packaging
Conference of Institute of Food Sciences and Technology, August 1974

Bureaux Internationaux Techniques (Brussels)
The recycling of plastic materials, September 1976

Business Monitor PQ276·1, PQ491, PQ492, PQ496
Business Statistics Office
HMSO, 1977

Business Monitor PQ271
Business Statistics Office
HMSO, 1977

European Plastics News **4** (1) pp12–34
Plastics Industry in Britain, 1976
IPC Business Press, January 1977

Cheater G et al
A study of reclamation and re-use of waste tyres
Rubber and Plastics Research Association, 1976

Dannenberg E M
Carbon black and the environment
Rubber Age **108** (4) pp37–41, April 1975

Data Base
The potential for energy conservation in nine selected industries
Gordian Associates Inc. (New York), 1974

Freshwater D C
Energy, economy, chemistry and industry
Society of Chemical Industry, November 1975

Gardener G
Machinery energy losses
Plastics and Rubber Weekly **699** p51, September 1977

Imperial Chemical Industries Ltd
The competitiveness of LDPE, PP and PVC after the 1973 oil crisis, 1974

Johnson G E
Energy consumption and drive selection for plastics pelletising machines
Society of Petroleum Engineers Conference (USA), Paper 012, 1976

Malaysian Rubber Producers Research Association
Private communication, 1977

Modern Plastics International
Even if resin is plentiful fillers will play a big role
McGraw-Hill, October 1975

Murakami K
Energy conservation techniques in plastics processing
Japan Plastics Age (Tokyo), pp29–34, January 1975

National Economic Development Organisation
The increased cost of energy—implication for UK industry
HMSO, 1974

Plastics and Rubber Research Institute TNO (Delft, Netherlands)

Battelle Institute (Switzerland)
Studie über den Beitrag abbanbarer Kunstoffe zur Lösung des Müllproblems, 1974

Makino H and Berry S
A thermodynamic analysis of packaging transport and storage
Illinois Institute for Environmental Quality, 1974

Midwest Research Institute (USA)
Report on plastics—resource and profile analyses, 1974

Sundstrom G
Beverage containers and energy
Rigello Pak AB (Sweden), 1974

TNO
Private communication—visit to bottle producer, 1974

TNO
Private communication—project information, 1974

TNO
Packaging as found in the Dutch market, 1974

US Department of Commerce
A feasibility demonstration based on the bottle manufacturing industry report
IRT 305–FR, 1974

Roger E R and Roebuck H S
Improved processing economics through scorch control
American Chemical Society (Cleveland), Paper 48, 1975

Rubber statistical Bulletin **31** (11)
International Rubber Study Group, 1977

Schaefer and Watson
Initial study of energy content of products and energy savings in their production
EEC (Brussels), 145–74–ECIC, 1975

Spencer-Palmer H J
Effects of increased energy costs on the competitiveness of plastics
British Plastics Federation, Symposium, 1974

Tunbridge T
Fillers and reinforcements
European Plastics News, pp45–49, October 1975

Tyre manufacturers
Private communication, 1977

Wolff S
Fillers—today and tomorrow
Deutsche Kautschuk Gesellschaft (Dusseldorf), September 1977

Wolk R H et al
The pyrolysis of tyres
Rubber Age (New York), p27, June 1974

Yokoyama Y
High speed two-stage extruders
Japan Plastics (Tokyo), pp13–18, September 1975

6 Glass and ceramics

Chapter contents

Glass

Total energy usage

Table 6·1 gives an overall view of the amount of glass manufactured in the UK from 1970 to 1975 and the energy used in its production. This table represents the usage of energy delivered to the factory, not the primary energy requirement.

Table 6·1
Glass production and energy consumption in the UK

Year	Glass production (tonne$\times 10^6$)	Energy consumption (GJ $\times 10^6$) Electricity	Oil	Gas	Solid fuel	Total (GJ/tonne)
1970	2·5	4·0	39·3	5·5	1·75	20·1
1971	2·6	4·1	42·4	7·3	1·42	20·9
1972	2·7	4·2	39·2	9·6	0·69	19·5
1973	3·1	4·3	37·4	11·2	–	17·3
1974	2·9	4·3	33·9	15·2	–	18·3
1975	2·8	4·2	25·9	15·1	–	16·4

The main indications are of a gradual shift from oil to gas to supply the main energy requirement, a sparing use of electricity, a phasing out of the use of solid fuel, and an improvement in the efficiency of fuel utilisation of about 20% over the period. The figures cover a wide variety of glass products. There are also variations within any sector of the industry so it is necessary to consider the industry by sector and by process operation to obtain a clearer view of the energy usage and the potential for conservation.

During 1975, the total manufacture of 2·8 million tonnes was distributed between the sectors:

flat glass—0·539 million tonnes
glass containers—1·825 million tonnes
domestic glassware—0·110 million tonnes
fibre glass—0·088 million tonnes
others—0·194 million tonnes.

Energy usage by process operation

In glass manufacture, a mixture of oxides together with some broken glass (cullet) is heated to a temperature of about 1 400 to 1 600°C to form a liquid melt. The melt is then refined to remove bubbles before being conditioned and fed to a variety of forming processes. The formed product is subjected to further processes (usually including annealing) which may involve significant energy consumption. The distribution of energy usage between the various processes for different glass products is given in Table 6·2. Separate figures are given for glass fibre insulation and reinforcement products, and those for lead crystal glass are based on a single factory.

The figures emphasise the energy intensive nature of glass melting. The melting units are all designed for continuous flow operation (see Figure 6·1) with the exception of lead crystal manufacture which is by pot melters as

Table 6·2
Energy consumption for different glass making processes

Operation	Energy consumption (GJ/tonne) Flat	Container	Domestic	Fibre Insulation	Fibre Reinforcement	Lead crystal	Tubing
Mixing	0·05	0·03		1·9	0·5		
Melting		10·02	15·7	15·2	18·6	38·7	12·2
Forehearth		0·97			10·8		
Forming	12·50 (Melting, Forehearth, Forming, Annealing)	0·5	5·1	17·7	1·4		6·0
Annealing		0·5				1·56	
Processing	1·50			3·3	4·0	0·93	
Services	0·15		2·4	3·8	1·6	3·12	3·2
Overall	14·20	13·00	23·3	43·3	37·8	45·75	21·8

illustrated in Figure 6·2. The lowest melting energy consumption in Table 6·2 is 10 GJ/tonne (for container glass) but, in a more detailed analysis, particular furnaces can achieve melting efficiencies of about 7 GJ/tonne. The theoretical requirement for a soda-lime-silica glass suitable for containers or flat glass is about 2·7 GJ/tonne.

Figure 6·1
Typical large glass melting tank

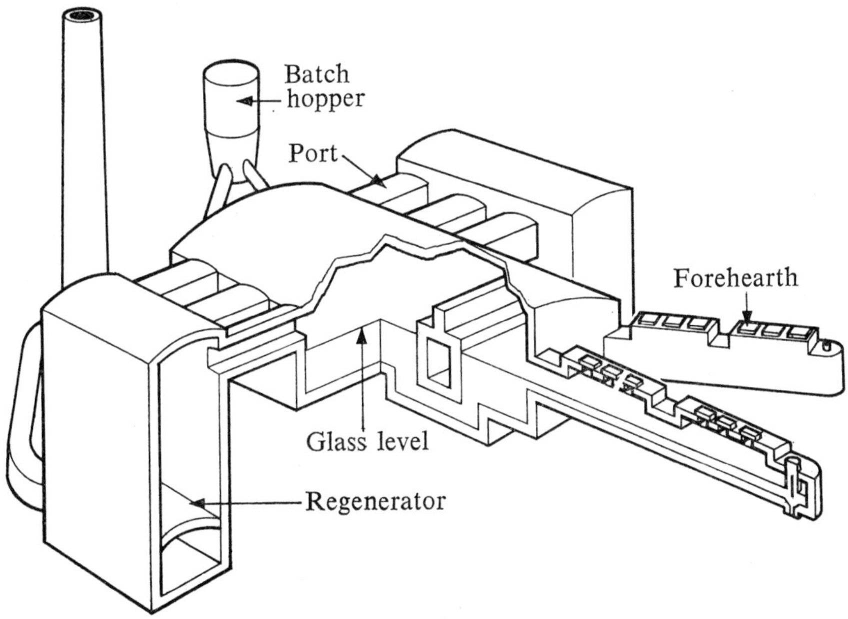

Figure 6·2
Multi-pot furnace

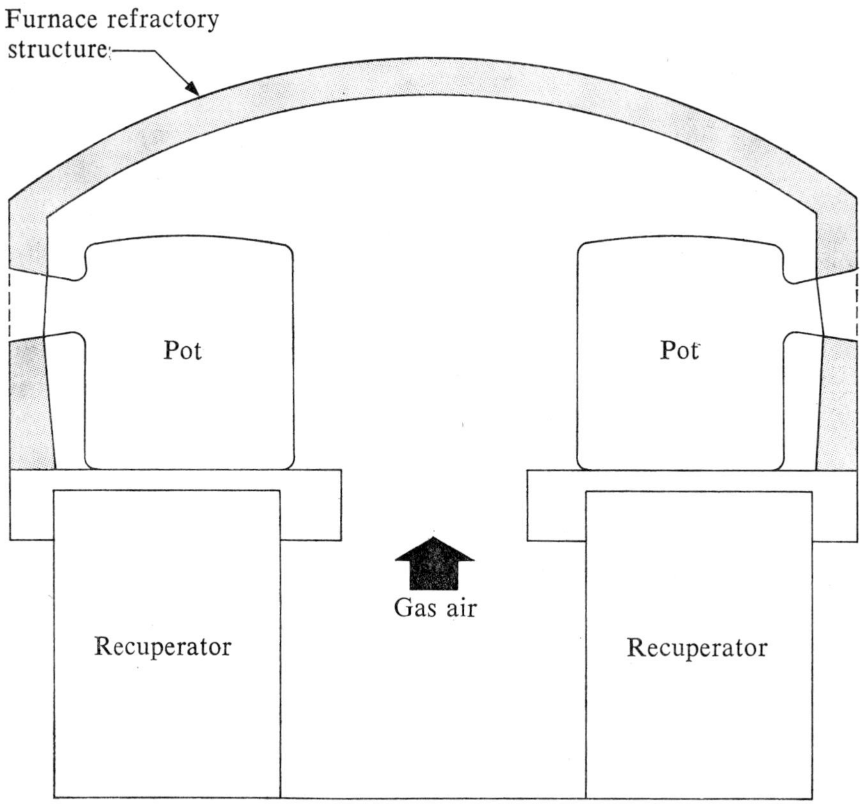

Replacing 1% of raw materials by broken glass reduces the fuel requirement by 0·2%. Where quantities of cullet are available from sources external to the factory and can be used without imparing glass quality, saving of energy and raw materials can therefore be achieved. This is generally true for the container industry but not for flat glass where the quality requirements are much more stringent and exclude the use of external sources of cullet.

The relatively high energy required to fiberise wool by high velocity hot gases after centrifuging is significant as is the energy required to control glass temperature prior to forming glass fibre for reinforcement or during forming of tubing.

Evolutionary technology for energy conservation
There are several obvious directions of technological evolution to improve fuel utilisation during melting. Since most furnaces are oil or gas fired the primary release of energy during combustion needs to be optimised by attention to burner designs and improved control of the combustion process. Regenerators or recuperators (for obtaining required pre-heat of combustion gas) need similar optimum design. Since 25 to 35% of the energy

input is rejected in waste gases, the use of waste heat boilers to recover a proportion of this high grade heat is receiving detailed attention. The use of heat from the waste gases to pre-heat the raw materials is a further possibility. Improved refractory materials, which enable higher melting temperatures to be used, also increase melting efficiencies by improving throughput and quality. The controlled use of insulation is being examined. There are particular circumstances within each part of the industry, for example particular glass quality requirements, which mean that the problems are not easy, but the broad direction of evolutionary progress is well defined. The Sankey diagrams (Figures 6·3 and 6·4) show the heat flows in typical float and container furnaces.

Figure 6·3
Typical float glass production Sankey diagram

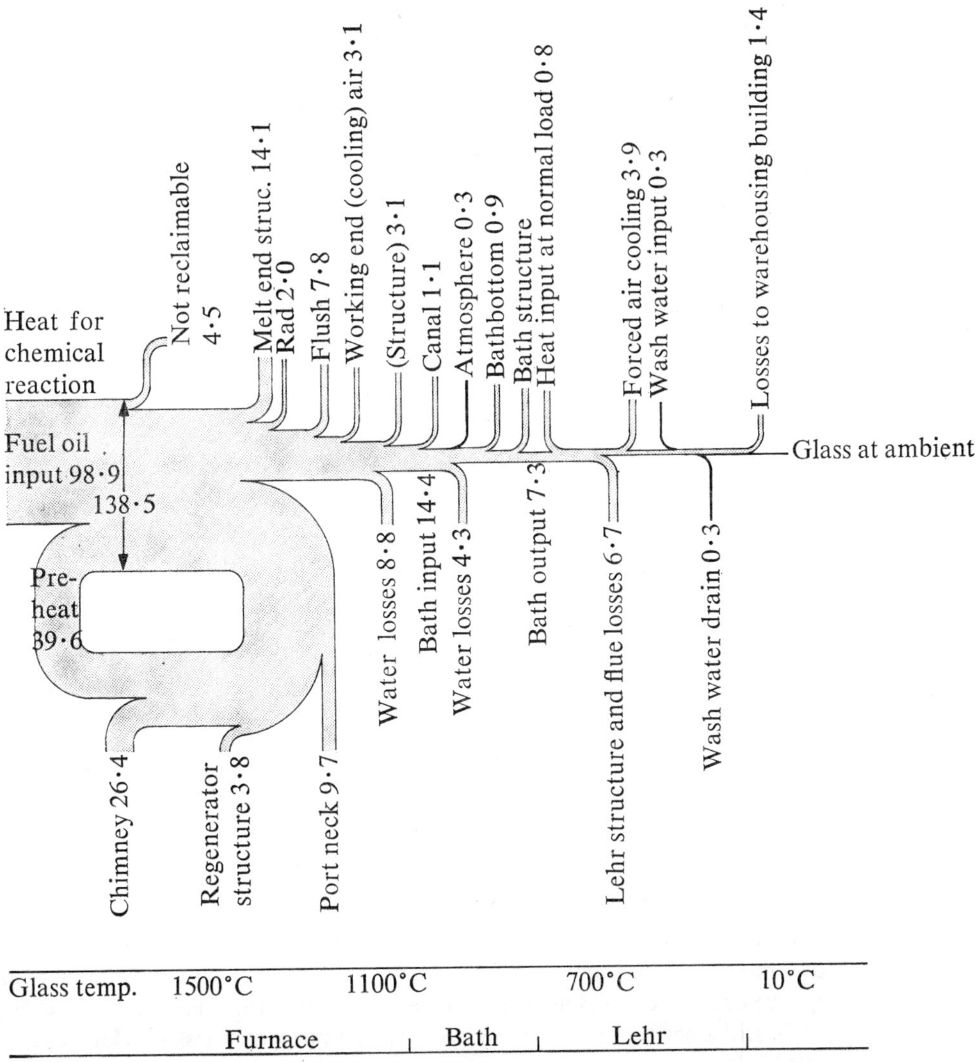

Figure 6·4
Typical medium-sized container furnace Sankey diagram

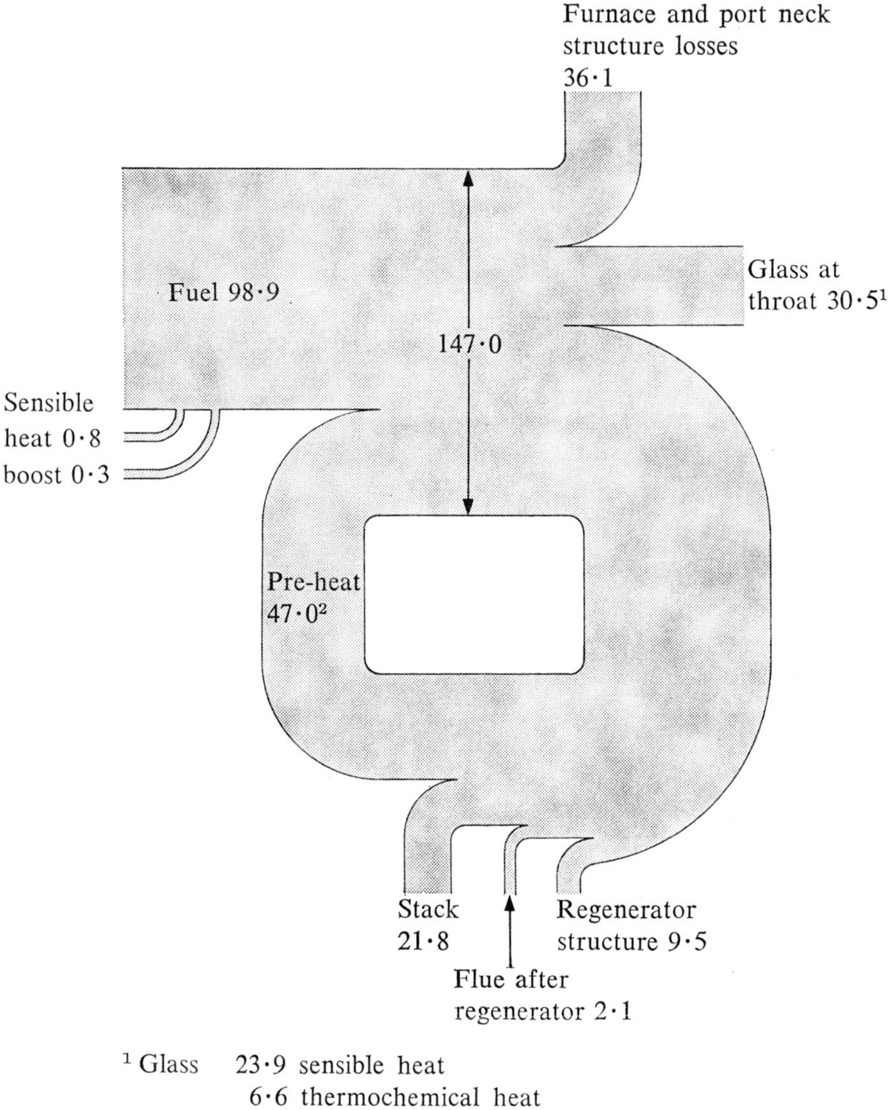

[1] Glass 23·9 sensible heat
6·6 thermochemical heat

[2] Preheat 43·3 from regenerator
3·7 flue before regenerator

It should be stated that the all-electric melting of glass by Joule heating of the melt is well known in the industry. The theoretical efficiency of energy utilisation during melting can be very high when electricity is used, but it is still uneconomic compared with the apparently less efficient use of oil or gas.

Revolutionary technology for energy conservation

A large proportion of the energy theoretically required to melt glass is needed to raise the raw materials from room temperature to about 1200°C, and the forming processes now used exploit the increase in viscosity of the glass as it cools. Thus low temperature (chemical) methods of glass formation would be unattractive unless they could include a forming as well as a melting operation. Nevertheless the challenge exists.

In the future, when gas and oil supplies diminish, the industry will again become dependent on solid fuel. The indirect use, via electricity or gas, poses no new fundamental problems. Is there, however, an elegant way of manufacturing glass using solid fuel directly?

Ceramics

Introduction

The ceramic products described in this section include refractories, bricks, pipes, whitewares, vitreous enamelling and some special newer materials. They all require energy to make, but the prime use of refractories is to save energy when producing other materials. Thus in this study it is essential to consider conservation both in the ceramics industry and in those industries which use ceramics. The relative sizes and energy requirements of various materials, ceramics and non-ceramics, is compared in Table 6·3.

Table 6·3
Energy requirements of refractories compared with other materials

	Annual production in UK (tonne×10³)	Energy requirements (GJ/tonne)	Total per year (GJ×10⁶)
Finished products			
Bricks	12 600	4	50
Refractories		2–200	34
Glass	2 500	24	60
Pottery		10–250	19
Intermediate materials			
Cement	17 000	13	220
Steel	20 000	45	900
Aluminium	350	230	80
Copper	70	45*	3
Plastic Polymers	2 000	90	180

The presentation of representative figures is difficult. For some manufacturing processes, such as brick making or the production of glass bottles, there are no intermediate stages and the end product only requires packaging and transportation. In most metallurgical industries, however, there is an intermediate stage such as an ingot or coiled sheet which requires further reprocessing by casting or pressing. The table has been divided to indicate such differences.

*45 is a maximum; the average figure is more like 15.

The other major problem associated with data of this type is the very wide range of energy requirements for the products of any one industry. For example, in the production of pottery the firing temperature, the number of firings and the number of reject pieces vary sufficiently to give the range quoted of 10–250 GJ per tonne.

In addition to these two general points, there are also the factors which affect individual products or materials. With most metals the energy cost is dependent upon the amount of recirculated scrap used and this will undoubtedly increase in the next few decades. The figure of 50 000 GJ per year given for brick production includes approximately 20 000 GJ of energy obtained from the carbon content of the clay used for their production. Considerable doubt must be expressed over the inclusion of this in the final total in view of the improbability of an alternative use for this carbonaceous material.

Despite all these qualifications the publication of such a table appears to be essential for the assessment of the current energy situation in the materials field and the distribution of research effort to those areas where the maximum benefit may be obtained. For this purpose figures indicating typical energy contents of materials and the total energy used by any one industry should be available.

Types of ceramics

Refractories

As refractory products are used almost entirely by other high temperature industries, changes in production are to some extent governed by changes in the user industries. The adoption of the Linz Donnawitz (LD) steelmaking process by the British Steel Corporation together with the changeover from coal gas to natural gas has led to a massive reduction in the production of silica bricks. This loss of production has been balanced by an increase in the other refractory materials, mainly high alumina and magnesite. The other major change has been in insulation materials. These were usually produced in the form of low density bricks from diatomite type materials and were placed behind the refractory material. The development of fibres capable of operating at temperatures over 1 200°C has led to their use as insulating blankets in place of the low density bricks. This has resulted in a considerable saving in energy and particularly a reduction in the heat capacity of kiln walls.

Conservation of the three raw materials used in the refractories industry (chromite, alumina and magnesia) cannot be done at the expense of the high temperature industries which require the end products. A thinner refractory brick will fail sooner and there appears little scope for redesign of these products.

Most effort is being made to improve the application of the materials with particular emphasis on the fibrous type of thermal insulation. This is occurring against a background of the continuous upgrading of many refractories by using materials of higher purity to reduce the variability of the product.

Bricks and pipes

Since the 1960s bricks have been fired in some form of continuous kilns, a system which brought an increasing amount of automation and a reduction of manual labour. It has been accompanied by attention to the raw material with the result that greater uniformity of product is now possible. Improvements have been achieved by more careful characterisation of the clay and, in some cases, by fairly sophisticated blending techniques. The design of bricks appears to have been finalised by the optimum distribution of the holes, and pipes by a strengthening of the fired material to enable wall thicknesses to be reduced.

With the use of waste heat for drying the bricks before firing, the only other improvement would be the use of heat from kiln exhaust gases. There is some work in progress on fast firing but so far no industrial use has been made of these techniques to reduce the firing time below 50 hours. The top temperature is governed by a variety of considerations; the need for specified crushing strength for engineering bricks or the development of a certain colour required by the consumer in the case of facing bricks. Greater use of low grade carbonaceous additives, such as domestic refuse, to the clay could further reduce fuel consumption.

Except for the application of fibre type insulation in the tunnel kilns there appears little scope in this branch of ceramics for saving either energy or materials, and the only matter worth more investigation is fast firing. Real savings could be made, however, if these relatively low energy products were used to replace metallic ones; clay pipes instead of steel pipes, and bricks instead of steel supported structures.

Whitewares

Whitewares include tableware, tiles, sanitary ware and electrical porcelain. This classification has little logical justification except that it covers most ceramic products excluded from refractories, bricks and pipes and special ceramics. Most of the products are used by the general public and in many cases their appearance is their major commercial attribute. For this reason some of the products are fired several times to obtain the desired appearance, for example in tableware or tiles.

Because of the wide variety of products, it is difficult to give a comprehensive outline of recent technical developments. With tableware, some mechanisation has occurred and reduction of the water content of the body by isostatic pressing is being investigated. Some tableware is now being produced with only one firing as opposed to the original two, one for the body and one for the glaze. Some investigation into drying has always been in progress, particularly where massive pieces such as wc pans are concerned. At least one firm made small tiles using a very fast firing process but these are no longer in production. Laboratory investigations into the saving of energy by reducing the temperature required for firing have also met with some success. Further investigation of the composition of pottery bodies might lead to a further slight reduction in the firing temperature required. A greater saving of energy might be expected from the application of the newer types of thermal insulation now available, the construction of kilns with lower heat capacities, and waste heat recovery.

Plentiful supply of most of the raw materials used and the present percentage utilisation gives little need for extensive materials conservation.

Vitreous enamel wares

Vitreous enamelling is the process by which glass is bonded to sheet steel to give a protective layer 0·1-mm thick. The method now used consists of the application of an aqueous suspension of glass frit to the metal component which is then dried and subsequently fired at about 800°C for approximately six minutes. This is the direct-on or one-coat process and it requires special steel with a carbon content less than 0·005%.

Originally two coats were applied, one dark in colour containing the adhesion oxides (oxides of nickel and cobalt) and a second coat to give the desired white colour. Within the last decade, application has changed from manual spraying to electrostatic spraying and work is now being done on both electrophoretic and dry electrostatic applications. The aim of these new methods is to produce an even coat of appropriate thickness with as little loss of frit as possible.

Traditionally, two types of enamel have been produced; one for the domestic market (largely cookers and cooking ware) and one for use in the chemical industry. The latter is resistant to both acids and alkalis but is specialised and expensive. More recently, other applications have been investigated and improved properties achieved either by altering the formulation of the frit or altering the surface properties. The development of the self-cleaning domestic oven was based on a catalytic material incorporated into porous enamel.

The major contribution to be expected from this branch of ceramics is in the protection of metals from corrosion. Present coatings can be applied to structural steelwork and new types of coating, possibly using some of the special ceramics, could be used for temperatures above 500°C.

Special ceramics

New inorganic materials have been introduced to the ceramics industry, including alumina, zirconia, silicon carbide and silicon nitride. They are often produced as high purity materials with outstanding properties such as very high melting point or very good thermal shock properties. Both alumina and zirconia are used in bulk as refractory materials but the important aspect of special ceramics is their potential to replace metals at high temperatures. The temperature of energy converters is governed by the alloys used in their construction and replacement by ceramics could lead to higher temperatures and increased thermal efficiencies.

Work in progress falls into three categories: studies of the production and general properties of the materials, investigations into their strength, and a limited amount of work on their application in energy converters. Work on the production methods now appears almost complete and methods are available for the fabrication of these materials in a wide variety of shapes.

The most promising area for research appears to be the collaboration of various disciplines to design, fabricate and test components made from these newer materials. Work is also required on other possible applications of inorganic materials at high temperatures such as their use in bearings and as lubricants.

Conservation

Conservation within the ceramics industry

The possibility of large-scale savings of valuable raw materials within this industry appears unlikely as most of the materials are both cheap and abundant. Production losses are particularly important in some sectors of the industry such as pottery and vitreous enamelling, and this problem is under constant review by the manufacturers. Saving of materials and energy by the reduction of these reject pieces does not appear possible by the application of general principles. The only recent major advance has been the redesign of clay pipes which led to an increase in strength and a reduction in wall thickness. This resulted in considerable cost savings. Other components, such as bricks, might benefit from similar study.

There are two important materials used by the refractories industry which are either scarce or energy rich—chromite and magnesia. As these are used exclusively for products concerned with the conservation of energy in the metals industries, there does not appear much scope for savings unless substitute materials can be found.

Most of the energy used in the ceramics industry is for drying and firing operations. These operations have been extensively investigated on a comparative basis by the British Ceramic Research Association and there seems to be only two areas where further work could possibly lead to energy conservation; the optimisation of firing schedules to achieve minimum use of fuel, and new body formulations to use lower firing temperatures.

Conservation using ceramic products

Refractory products are designed for containing processes involving high temperature and corrosive materials or for preventing heat losses from these processes. Maximum returns will be achieved if refractories are used most efficiently in those industries which use most energy. This involves first paying attention to the highest temperature processes (around 1 800°C) and then to the lower temperatures (about 1 000°C) as found in brick kilns. The major factor here is the new fibrous type of insulation and its use should be investigated at both the hot and cold faces of the kiln wall.

The second most important area for conservation is the substitution of ceramics for metals or plastics. This would save both energy and scarce raw materials. On a weight basis, steel requires about five times the energy for manufacture of most ceramics and the relative abundances of the raw materials must differ by a factor greater than five. Comparisons are shown in Table 6·4.

Table 6·4
Energy of manufacture and reserves of raw materials

Material	Energy[1] (MWh/tonne)	Reserves[2] (years)
Aluminium	64	100
Plastics	25[3]	31[4]
Steel	15	240
Glass	6	1 000[5]
Ceramics	3	1 000[5]

[1] Based on: Material Needs and the Environment Today and Tomorrow, Final Report. National Commission on Materials Policy (USA), 1973.

[2] Based on: Meadows D H et al. Limits to Growth. Earth Island Ltd. 1972.

[3] This includes the energy value of the material itself, 13 MWh/tonne.

[4] If plastics are manufactured from coal as well as oil, this figure should be much larger.

[5] Arbitrary figure for reserves of silica and silicates.

There are doubts concerning this type of information. Firstly, the associated energy requirements such as transport and installation are not taken into account and, secondly, the figures will be outdated as improvements in production and recycling occur. However, in most cases the associated energy costs are only a small proportion of the whole and the costs are likely to be similar when alternative materials are used. Future energy costs are far more relevant but, bearing in mind that future changes can only be estimated, the current figures given are the best available. Such data of course must be used correctly and differences of less than a factor of 2 are probably of little significance.

More work should be done on the use of engineering ceramics, especially where the increased temperatures attainable would result in an increased fuel efficiency. Attention should also be paid to the less sophisticated products such as the use of materials based on clay in structural applications or clay pipes instead of plastic pipes. The choice of constructional materials is very complex, sometimes governed by tradition, sometimes by cost, but a systematic investigation of substitution should be made.

Ceramic coatings could be used more widely for the protection of metals at both ambient and high temperatures. There should be an examination of coatings of some of the newer ceramics for their properties as bearing surfaces at high temperatures.

University research

Glass

It is highly desirable to encourage direct interaction between members of the academic community and the glass industry in order to understand correctly the problems and the opportunities for conservation. Perhaps this is particularly needed in smaller organisations which do not have the manpower resources to allow a full concentration of effort on these problems. Also, any potential solutions must be placed against the background of commercial operation of the industry in order to make a realistic contribution.

Possible subjects for university research are listed below.

- Develop useful ways of conserving the considerable amounts of low-grade heat (for example, cooling water at 50°C) rejected by glass-making processes.
- Develop higher quality refractory materials for furnace superstructure, regenerators and glass contact to improve both fuel utilisation and glass quality.
- Improve understanding of the basic melting and refining operations in order to design more efficient furnaces and control them more closely. The ability to predict flow and temperature regimes within the glass in novel designs of melter would be desirable. Present studies concentrate on modelling existing systems.
- Examine low temperature routes to glass product manufacture.
- Investigate the possibility of producing glass directly from a solid fuel energy source.

Ceramics

Possible subjects for university research are listed below.

High temperature processes

- Study the thermal behaviour of kiln walls with various types of construction.
- Match kiln wall structure to the firing cycle and to the production schedules of specific products.
- Investigate the stability of fibrous insulation materials in various environments.
- Study fast firing techniques.
- Examine the changes needed to body compositions and raw materials to permit lower firing temperatures.

Substitute materials

- Extend the engineering uses of the newer ceramics.
- Investigate systematically the development of low temperature bonds and desired microstructures in ceramic materials.
- Investigate the fabrication of ceramic materials suitable for making large sheets.
- Develop new and cheap ceramic coatings and coating methods.

7 Cement

Chapter contents

Introduction

An awareness of the costs of fuel and power is common to virtually all cement makers but the level of interest in economy measures has varied widely over the years from country to country depending upon the relative costs of fuel and power and on the skilled labour used to control and maintain high performance.

Manufacturers in continental Europe have been particularly fuel conscious with the result that over 80% of the cement is made by dry or semi-dry process plants. In Britain, the presence of soft wet raw materials with hard flint inclusions (which can most readily be extracted by slurrying) and the availability of relatively cheap coal have combined to keep the wet process for 69% of cement production. However, where the opportunity has arisen, dry or semi-dry process plants have been introduced over the past 20 years. The effect of this and continued effort to improve the fuel consumption of all plants has resulted in a fall in fuel consumption over the years as shown in Figure 7·1.

Cheap oil during the 1950s resulted in a number of conversions from coal at the works near to the coast but, with the sudden upsurge of oil prices in recent years and the closure of the older plants, the largest UK manufacturer now has only two oil fired units. Both are distant from the coal fields.

Figure 7·1
Kiln fuel consumption for Blue Circle Industries Ltd (all processes)

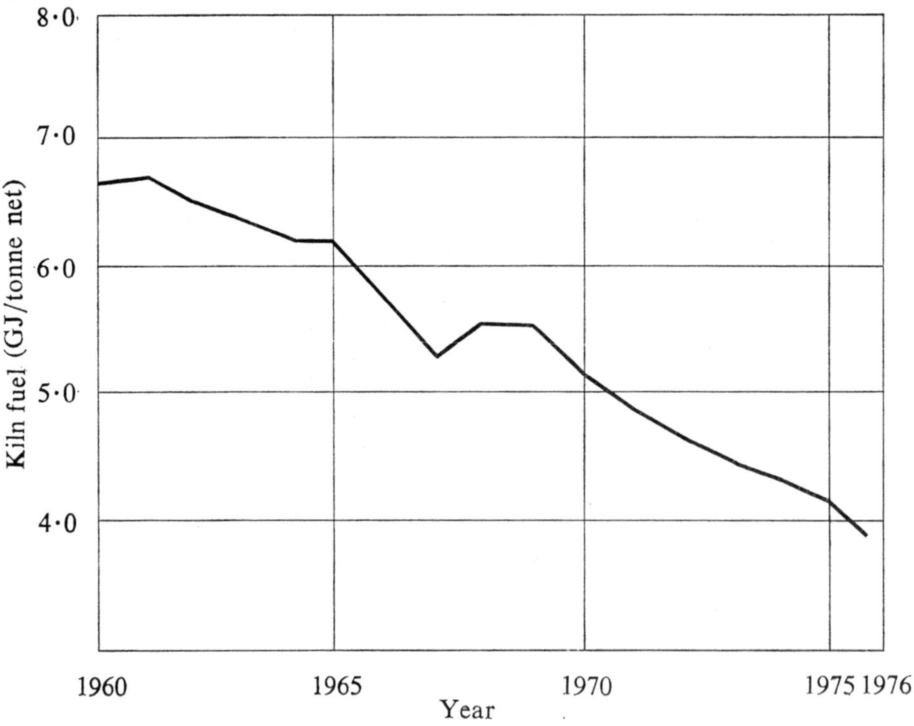

Figure 7·2
Relative UK coal and power prices

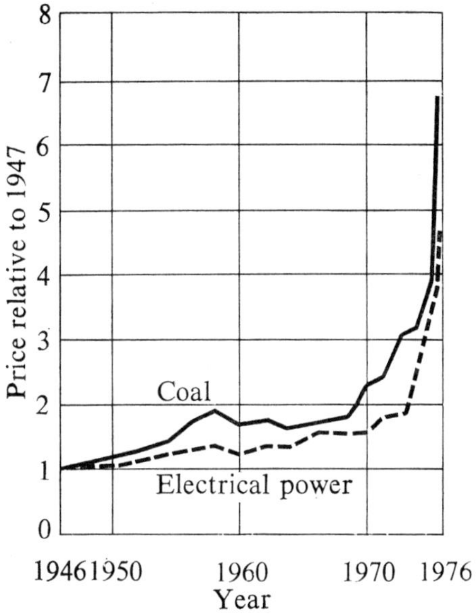

Figure 7·3
Relative UK coal and power prices, adjusted for inflation

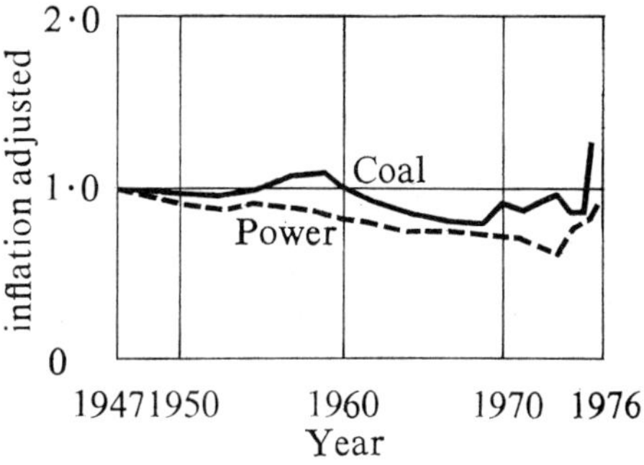

Since 1973, the prices of all sources of energy have increased dramatically as shown in Figure 7·2. The result is that manufacturers have studied the changes, if any, that should be made to the wet process and other fuel economies that can be introduced to each plant, wet or dry.

One outcome of the studies is that a great deal depends upon future cement requirements—which at the present time are not encouraging in the UK. In a static market, the costs of converting from wet to dry process are not usually justified despite the fuel price rises.

Therefore, while there is an almost universal outcry to save fuel, the basic economics do not support widespread further conversion at present from wet to dry process in Britain. Some support for this can be seen in Figure 7·3, which shows the relative fuel and power prices of Figure 7·2 adjusted for UK inflation. Only since 1976 have the fuel prices exceeded the position they were at relative to the general commodity price level in 1947. In a steeply inflationary situation high rises in fuel prices have to be weighed in relation to increased costs of the machinery used for process conversions. This situation may differ in other countries.

This chapter explores all possible ways of improving the performance of the existing plants and the whole energy side of the industry. It is based on papers prepared for use within Blue Circle Industries Ltd.

Analysis of present energy usage

Table 7·1 compares the energy used for cement production and delivery and in the various administrative offices, depots etc. for one company, based on 1974 data. The energy to provide kiln fuel (mainly coal) represents 91% of total energy used, but only about 63% of the total energy cost. Electrical power at 6% of total energy used represents about 31% of energy costs and transport fuel but at about 3% of total energy it represents 6% of energy costs.

Table 7·1
UK energy consumption for Blue Circle Industries Ltd, 1974

Section of industry	Coal-heating and burning (% of total)	Oil-transport (% of total)	Electricity (% of total)	Total (%)
Administrative establishments	0·04	–	0·01	0·05
Cement delivery	0·005	1·5[1]	0·04	1·5
Cement production	91·1	1·2	6·1	98·4
Total consumption	91·1	2·7	6·1	100·0
Estimated fuel costs	63·2	6·0	30·8	100·0

[1] Represents road delivery by cement lorries only.

The actual amounts of fuel and electrical power used vary with the type of process, the raw materials used, the size of the plant and its general performance. Table 7·2 gives a guide to the energy and power figures from the better plants.

Table 7·2
Fuel and power consumption at efficient cement plants

Plant process	Fuel (net GJ/tonne of clinker)	Power (kWh/tonne of cement)
Wet process[1]	5·45	77
Lepol (including raw[2] material drying)	3·45	100
Suspension preheater[3]	3·15	120

[1] The raw materials are fed to the kiln as a slurry with a moisture content usually between 30 and 40%, measured on the wet basis.

[2] The raw materials are fed to the grate-kiln system as nodules of about 12% moisture content.

[3] The raw materials are fed to the cyclone preheater kiln system as a dry powder.

Factors influencing kiln plant energy consumption

Basis for kiln system studied

Figure 7·4 illustrates some of the major parameters affecting the fuel consumption of a wet kiln system. Similar factors apply to the dry suspension preheater process although their magnitude and effect are different. These quantitative effects on fuel consumption for both types of process have been investigated by calculating system heat balances. The results of these calculations are discussed in the following sections using, where appropriate,

Figure 7·4
Some factors affecting wet process kiln fuel consumption

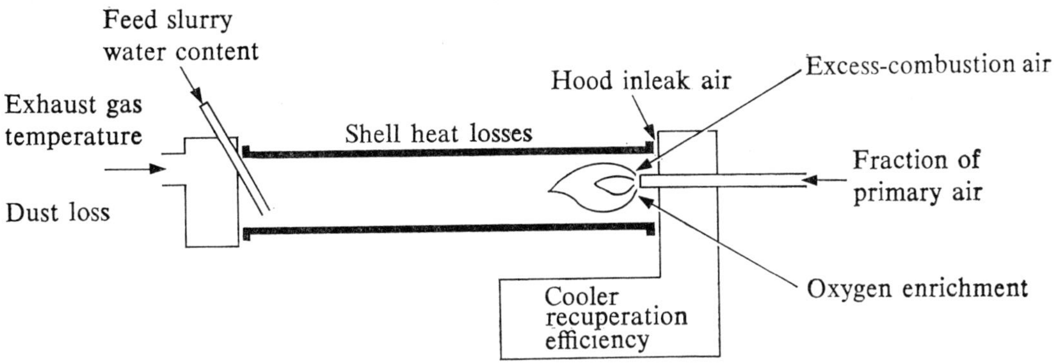

calculated gas and feed temperature distributions through the system.

The procedure has been to set up base level values for the parameters investigated and to consider the effect on fuel consumption of varying each parameter while holding all others constant. Base level values used and ranges of variation investigated are given in Table 7·3 and marked with asterisks on subsequent graphs.

This method does not provide a full mathematical model of the system because, in practice, alteration of any one parameter may also cause a change in one or more of the others. In this study it was assumed that suitable changes are made to the plant to enable the base conditions to be maintained. This is not in fact as artificial as it may first appear because, for example, a reduction in slurry moisture would normally raise kiln exit temperature and would be countered by adjustments to the kiln chains.

Kiln system thermodynamic performance

Figure 7·5 shows calculated gas and feed temperature variations through the system for a wet process kiln, plotted against cumulative heat transferred to the feed. The base conditions are those of Table 7·3.

The horizontal axis also represents approximate distance along the kiln or feed time of passage through the system. It has been assumed that the feed is heated to 67°C before any drying takes place and that the drying takes place uniformly between 67 and 107°C feed temperatures. The feed is then heated to 558°C and all calcining takes place uniformly between 558 and 1 060°C. The general assumption is that feed composition is a function of feed temperature only.

Table 7·3
Conditions for calculations

Parameter	Wet process		Dry process	
	Base condition	Range investigated	Base condition	Range investigated
Back-end temperature (°C)	200	150–400	–	–
Minimum gas-feed temperature difference, ΔT (°C)	–	–	120	0–400
Cooler recuperation efficiency (%)	69	0–100	48	0–100
Primary air (% of total combustion air)	25	0–40	25	0–40
Hood inleak air (% of total combustion air)	10	0–60	10	0–60
Shell losses (MJ/tonne clinker)	840	0–1 700	420	0–1 700
Excess air (% of stoichiometric combustion air)	5	0–40	5	0–40
Oxygen enrichment (oxygen addition/oxygen in secondary air) (%)	–	0–50	–	0–50
Dust loss (% on clinker)	10	0–60	–	–
Slurry moisture (% water in slurry)	35	20–40	–	–
Fuel consumption calculated (net GJ/tonne clinker)	6·16	–	3·31	–

Gas temperatures through the system have been calculated by starting at the back-end of the kiln and working through the kiln, assuming a counter-flow situation, taking account of heat transferred to the feed and of kiln shell losses. A further condition in the calculation (chosen for the base condition) is that the kiln gases leave the kiln at 200°C.

It can be seen that the gas temperature curve is approximately linear. Deviations from linearity are caused by variation in shell losses and gas specific heats and by the mixing in of cooler gases evolved from the feed. When comparing the gas and feed temperature curves with the real situation the temperatures must be regarded as cross-sectional averages. Gas temperatures towards the hot end of the kiln will differ from the model, varying with flame length and also being affected by changes in the specific heat of the gases at very high temperatures and under conditions of some dissociation. These deviations from the model system are not important for this study which is primarily concerned with the drying and calcining zones of the kilns and with overall heat balances rather than the burning zone.

Because the heat transfer rate is dependent on the temperature differential, ΔT, of the gas-feed, it is important when investigating the thermodynamic performance of a kiln system to look at where this temperature difference is low. For the wet process kiln (Figure 7·5) there are three minima of ΔT. They are at feed entry to the kiln (ΔT=180°C), at the commencement of calcining (feed temperature 558°C, ΔT=628°C) and at maximum feed temperature (1 500°C, ΔT=881°C). Because the gas-feed temperature difference at feed entry to the kiln is by far the lowest, it is reasonable to use this temperature difference (or the kiln back-end temperature) to characterise the kiln performance. This is commonly done in practice where we look at kiln back-end temperature to see how well the kiln is performing thermally.

The gas and feed temperature curves for the dry process kiln system are considerably different to those of the wet system. Figure 7·6 shows the situation for a true counter-flow arrangement (as in a long dry kiln) but this is approached in a multi-stage preheater system with a series of stepped counter-flow increments. In particular, the region of minimum temperature difference between gas and feed occurs at the beginning of calcination

Figure 7·5
Calculated gas and feed temperatures (wet process)

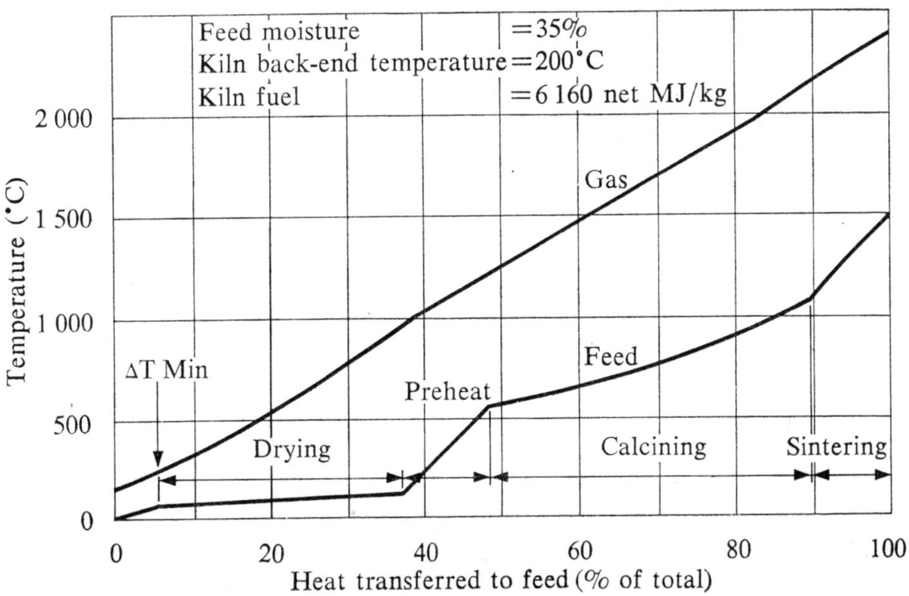

(assumed to be 700°C feed temperature for the dry suspension preheater process), not at entry to the system. This is an interesting point to emerge as it indicates that in order to assess the thermodynamic performance of a dry suspension preheater (SP) system we need to know this minimum temperature difference rather than the difference at the preheater exhaust which might be thought to be the important factor as in the wet process kiln. However, as with the wet process, the lower the minimum temperature difference, the better the system is operating.

Figure 7·6
Calculated gas and feed temperatures (dry process)

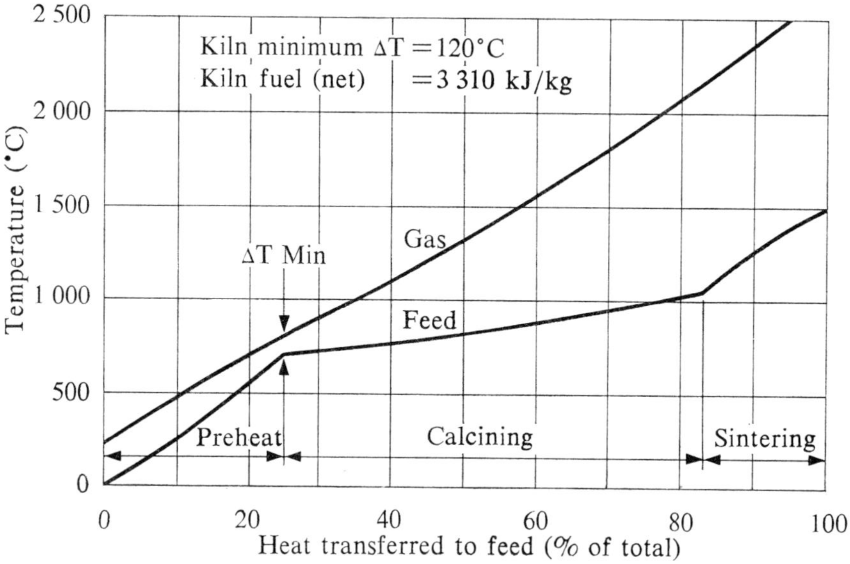

Effect of minimum temperature difference on fuel consumption
Wet process
The effect of kiln back-end temperature on fuel consumption is shown in Figure 7·7. A change of 100°C causes a change in fuel consumption of 500 net MJ/tonne of clinker. This change is less than might be calculated from a heat balance on the kiln alone because a cooler air flow affects the cooler recuperative efficiency (cooler performance is considered in greater detail below).

Figure 7·8 shows how the back-end temperature changes affect gas temperature distributions through the system. It shows that an increase of 200°C in the back-end temperature produces a change of about 120°C in the gas temperature at the start of the calcining zone and slightly reduces gas temperatures in the burning zone (due to the increased fuel consumption causing a drop in secondary air preheat temperature). Hence the proportional rise in gas-feed temperature difference is by far the greatest in the drying zone and small towards the hot end of the kiln. The converse of this is that the kiln exhaust gas temperature is mainly affected by the

Figure 7·7
Effect of back-end temperature on fuel consumption (wet process)

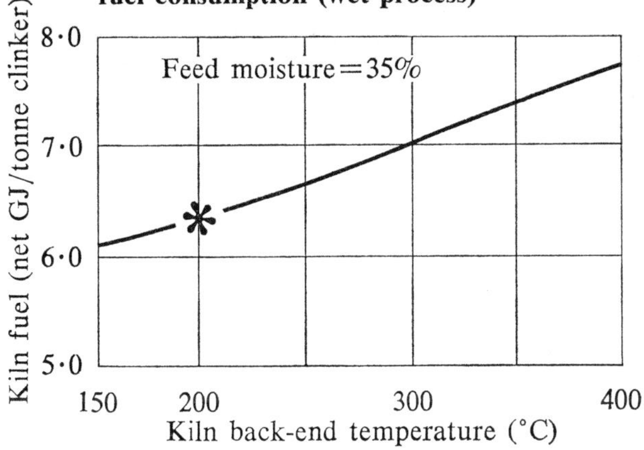

Figure 7·8
Effect of kiln back-end temperature (wet process) on calculated gas and feed temperatures

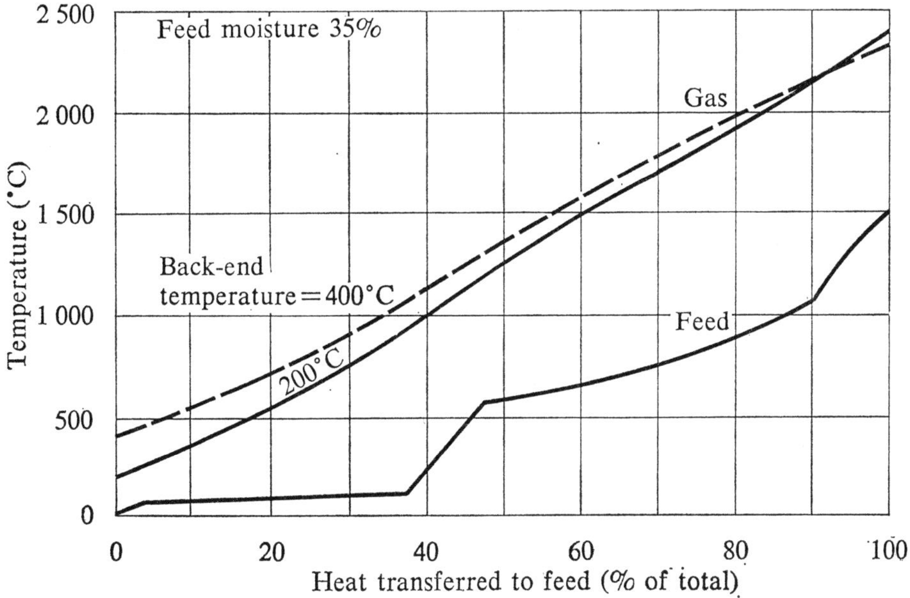

heat transfer performance of the kiln drying zone, i.e. in the kiln chains. The importance of correct chaining to keep exhaust temperature down and hence reduce fuel consumption is clearly shown in Figure 7·7.

It is not possible to go into chain design in any detail here although it is a topic of considerable interest to kiln men and a topic of some controversy. As a general policy, the chaining system adopted by the major UK manufacturer has been to instal evenly spaced spiral curtains spread over a rather greater length of kiln than commonly used. This type of chaining is exclusively fitted in 62 of their kilns in the UK and overseas.

This system should provide a slow even rate of drying which has been found to result in even material transport through the chains and a nodular product leaving them. A second and important objective is to provide a system which is relatively simple to instal and maintain. The design of curtains is straightforward since there are no complications of calculating garland lengths, angles of hanging, etc. Installation involves fixing to the attachments with only a single bolt. The decision as to how much chain to instal is normally based on a chain/clinker weight ratio which is chosen to take account of the kiln length/diameter ratio, the slurry moisture content and the physical properties of the raw materials (particularly friability and nodule strength). The performance of a kiln with a new chain system is closely monitored and if necessary the loading is adjusted to reduce the kiln exhaust gas temperature.

Dry process

The choice of a value for minimum temperature difference and its significance are not immediately obvious. The multi-stage suspension preheater (SP) is a stepped counter-flow preheater with the heat transfer performance (represented by minimum temperature difference between gas and feed)

Figure 7·9
Effect of gas-feed temperature difference on fuel consumption (dry process)

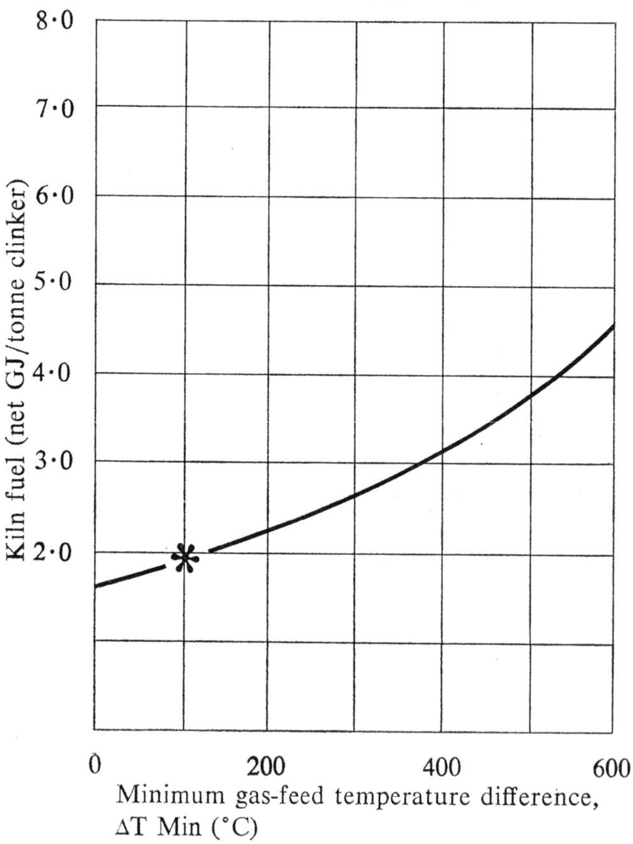

dependent on the number of preheater stages and the total work done in the preheater. A reasonable assumption is that the minimum temperature difference is approximately equal to the minimum feed temperature rise across any stage.

Figure 7·9 shows the effect of minimum temperature difference on fuel consumption. For a dry process kiln of infinite length or an infinite number of perfect preheater stages, the gas-feed minimum temperature difference will be zero. This gives a best possible fuel consumption of 3·01 net GJ/tonne clinker for the assumed base conditions. This is not the absolute minimum because there is still, theoretically, room for economy on the base level conditions.

This graph clearly shows the diminishing returns for increasing the number of preheater stages. For example, a typical minimum temperature difference between gas and solid in a four-stage preheater would be 120°C (shown asterisked on Figure 7·9) with a fuel consumption of 3·31 net GJ/tonne. Increasing the number of preheater stages would cause the minimum temperature difference to fall, the limit being reached at 0°C difference. The fuel consumption would then be 3·01 net GJ/tonne (all other things being equal), i.e. a saving of only 300 net MJ/tonne.

Figure 7·10 shows how higher minimum temperature difference increases exhaust gas temperatures. The figure highlights the case of a long, dry kiln where poor heat transfer in the kiln would be expected to raise the temperature difference to a level approaching that typical at this stage of a wet process kiln, i.e. 400 to 600°C. High fuel consumption will result.

Figure 7·10
Effect of gas-feed minimum temperature difference on calculated gas and feed temperatures (dry process)

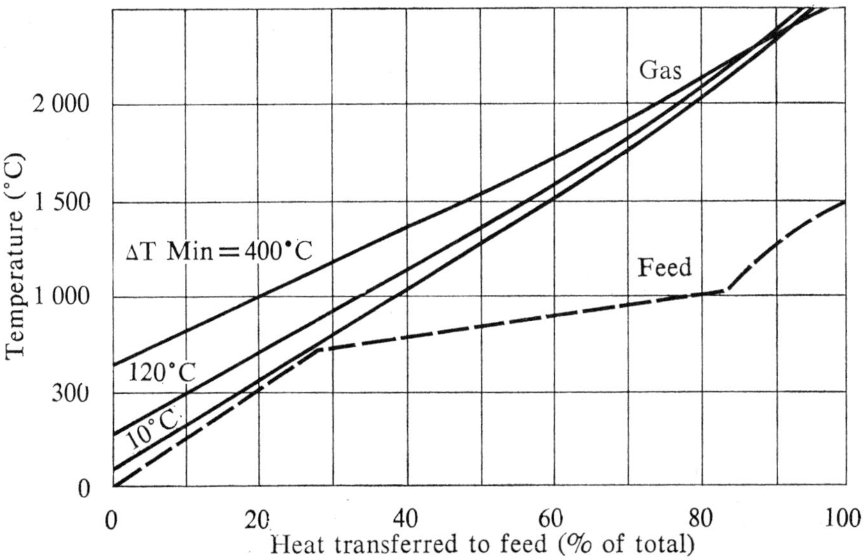

Overall these concepts of minimum temperature difference provide a picture of what is controlling the thermal process in these kiln systems. In the wet process, improved economy results directly from reducing kiln exit gas temperature. In the dry process, the preheater is the controlling feature. These concepts show there is not much to be gained by adding further preheater stages and show the factors that control the exhaust temperature. Economies have to be sought elsewhere.

Cooler recuperation efficiency

The cooler recuperation efficiency, η_c, is defined by

$$\eta_c = \frac{\text{heat recovered from clinker in kiln secondary air}}{\text{heat available in clinker leaving the kiln}}$$

A 10% increase in cooler efficiency can save about 140 MJ/tonne clinker for a wet process kiln and about 200 MJ/tonne clinker for a dry process kiln (see Figure 7·11). The precise savings available depend on kiln hood inleak air and primary air flows.

For a dry process kiln these factors also affect the maximum efficiency possible (about 69% for the case studied) which is reached when the gas leaves the cooler at the clinker exit temperature from the kiln. Another effect of improved cooler performance is that the gas temperatures towards the hot end of the kiln will increase with possible further reductions in kiln back-end temperatures. In practice, cooler efficiency increases with the secondary air flow through it. To take this into account, a theoretical

Figure 7·11
Effect of cooler efficiency on fuel consumption

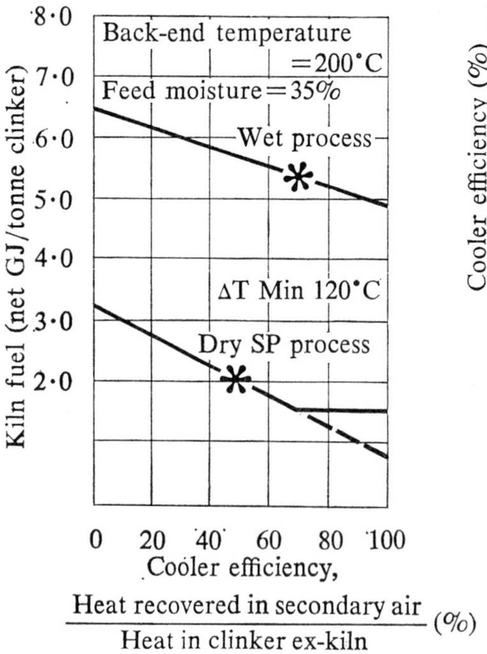

Figure 7·12
Grate cooler recuperation efficiency

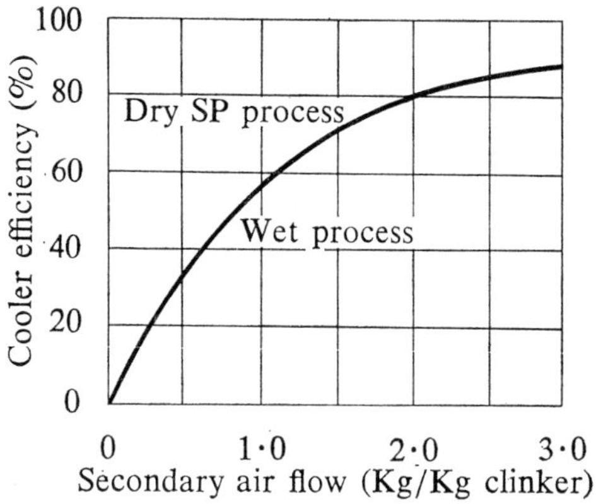

curve for a grate cooler has been calculated to show cooler efficiency as a function of secondary air flow (see Figure 7·12). This relationship has been used for all calculations (except those for varying cooler efficiency). The changing cooler performance has considerable effect on all the heat balances because any change that increases kiln fuel brings with it increased cooler air flow and improved cooler heat recuperation which acts to counter the increased fuel requirement. A similar situation arises when changes reduce kiln fuel requirement.

Other types of cooler behave in a similar way and, with all of them, the need to maintain efficiency when kiln fuel consumption is decreased by other economies will be apparent. This calls for higher temperatures in secondary air which in turn requires better materials of construction especially at the throats of planetary coolers and for lifters in both planetary and rotary coolers. Development work in these topics has been done and more is in progress.

Slurry Moisture

Slurry moisture is the key factor determining the fuel consumption of wet process kilns and its effect is shown in Figure 7·13. As an example it will be seen that a reduction from 40 to 30% moisture would save 960 net MJ/tonne provided the kiln chains could be suitably adjusted to hold constant the kiln exhaust temperature. Reductions of slurry moisture to 30% or below, from generally accepted levels around 40%, have been achieved at a number of UK plants in recent years. Two examples are shown in Figure 7·14.

Figure 7·13
Effect of feed slurry moisture on fuel consumption (wet process)

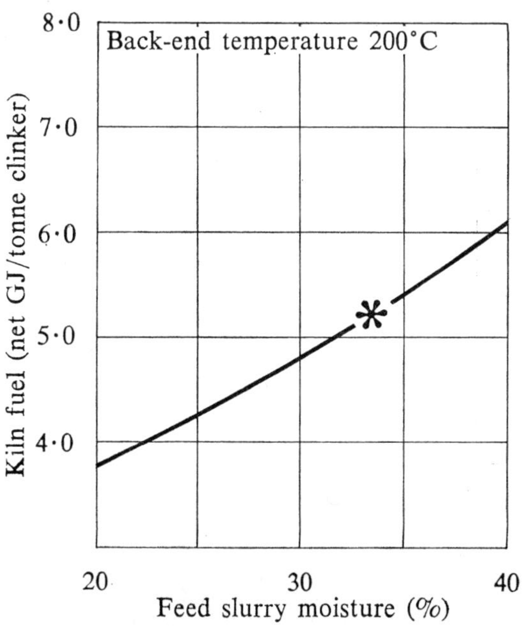

Figure 7·14
Reduction of feed slurry moisture at two UK cement plants

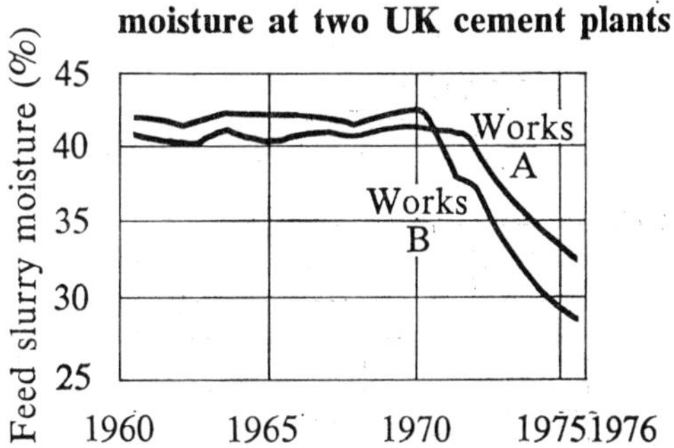

This is a profitable area for fuel economy but the situation and materials at each works have to be studied individually in a two part programme. The first part involves a careful examination of the existing system to find and overcome bottlenecks which prevent thicker slurries being handled. Inadequately angled troughs or pipelines with many acute bends may be the cause. Above all, it is vital for the plant operatives to be conscious of the need to reduce moisture and cut out practices such as adding water from hoses to overcome troublesome flow areas. Savings of 2 to 3% moisture can be made from these studies alone. The second part of the programme concerns the complex subject of chemical additives. Briefly, this involves consideration of the clay minerals present in the slurries, their behaviour under shearing conditions (as in pumps) and an examination of the water purity. Many raw material problems have been solved but a few difficult ones are still being investigated so that the downward trend in slurry moisture will continue.

For these studies to succeed it is important to recognise that dense, low moisture slurries will have certain unavoidable side-effects. Any classifiers in the raw mill circuit will fall off in performance, probably increasing milling power needs whilst chaining modifications will be required in the kilns.

Dust loss

As with slurry moisture, dust loss is essentially a wet process concern. Most dry process dust is automatically recycled in the raw meals. Figure 7·15 shows the effect on fuel consumption of dust lost and dust returned. Obviously the lowest possible dust quantity leaving the kiln is to be preferred but if it is caught and returned the small sensible heat loss is negligible compared with the losses if the dust is dumped. A 10% total loss of dust would cost 400 MJ/tonne in fuel consumption.

Figure 7·15
Effect of dust loss from the kiln on fuel consumption (wet process)

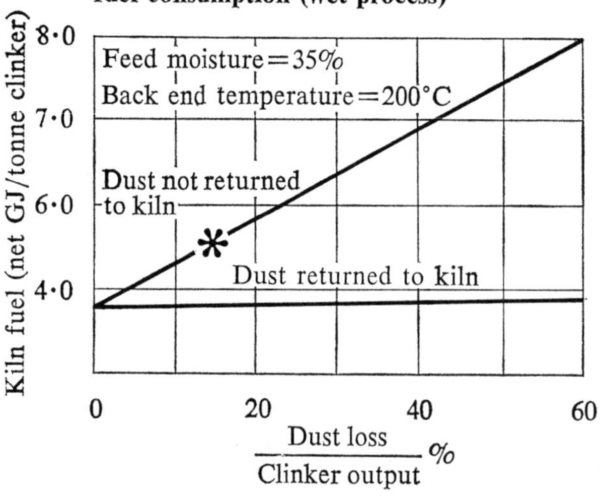

Various methods of dust return are employed but direct insufflation into the kiln burning zone is popular. Where dust is highly alkaline, direct return is not practicable and alternative routes have to be sought. A hot fluid bed system for extracting the alkali and leaving a clinker which can be added to the kiln product is being developed. This preserves the thermal energy put into dust instead of discarding it.

Primary air and kiln hood inleak air

Primary air and kiln hood inleak air both represent a loss of preheated cooler air and are analogous to a loss in cooler efficiency. Reducing inleak air (at 20°C) from 10% of combustion air to zero saves about 40 MJ/tonne clinker for the wet process kiln and 80 MJ/tonne clinker for the dry SP process kiln (see Figure 7·16). Reducing primary air (at 100°C) from 25 to 15% of total combustion air saves about 30 MJ/tonne clinker for a wet process kiln and about 80 MJ/tonne clinker for the dry kiln (see Figure 7·17). In all cases, the effects are less than given by a heat balance, assuming constant secondary air preheat temperature, because this increases as the secondary air mass flow drops. If the cooler on a dry process kiln is operating near maximum possible efficiency, then primary and hood inleak air reduction would bring about a greater reduction in fuel consumption than that given above because the secondary air temperature would be near

Figure 7·16
Effect of hood inleak on fuel consumption

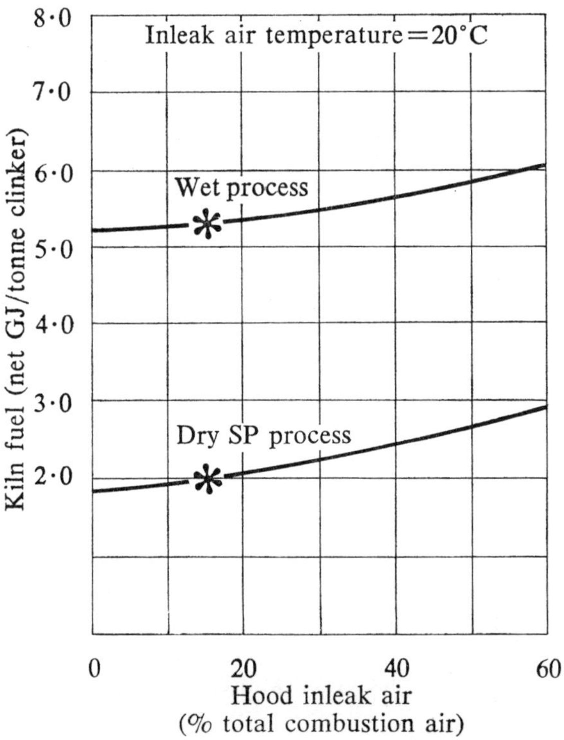

Figure 7·17
Effect of primary air on fuel consumption

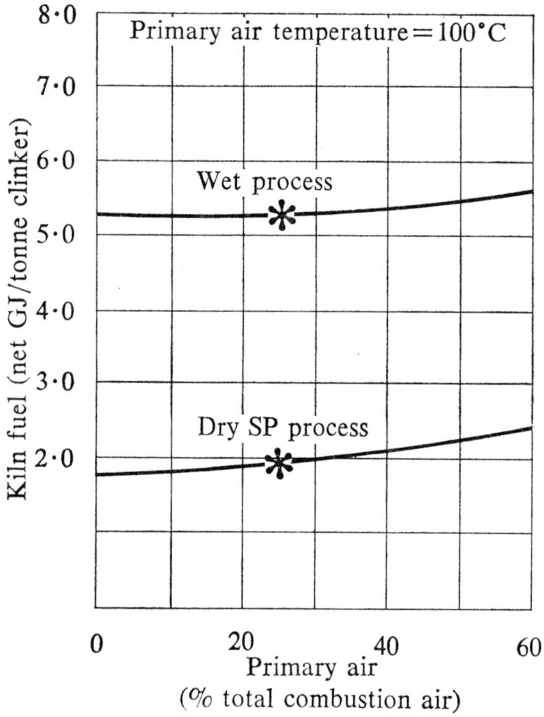

the clinker temperature leaving the kiln and would not drop so much with increasing mass flow.

Good maintenance can minimise inleak air but primary air on coal-fired systems is to a large extent governed by the firing system. With direct firing, straight from coal mill to kiln without any fine coal storage, the primary air is also the conveying air through the coal mill and is in the range 1·5 to 2·0 kg of air per kg of coal. For normal coals this fixes the primary air in the range 15 to 20% of total combustion air. Indirect firing using powdered coal from a storage hopper can be done with 10% primary air since it is only required to entrain the coal. While attractive from a thermal economy point of view, indirect firing introduces extra power costs and practical problems of collecting and storing fine coal.

Shell losses

The effect of kiln shell heat losses on fuel consumption are shown in Figure 7·18. The calculations made for a wet process kiln show that a decrease in shell losses of 400 MJ/tonne clinker produces a decrease in kiln fuel consumption of almost exactly 400 MJ/tonne clinker, with constant back-end temperature. This is at first surprising, because it might be expected that the decreased fuel consumption would bring with it decreased kiln exhaust gas

Figure 7·18
Effect of shell losses on fuel consumption

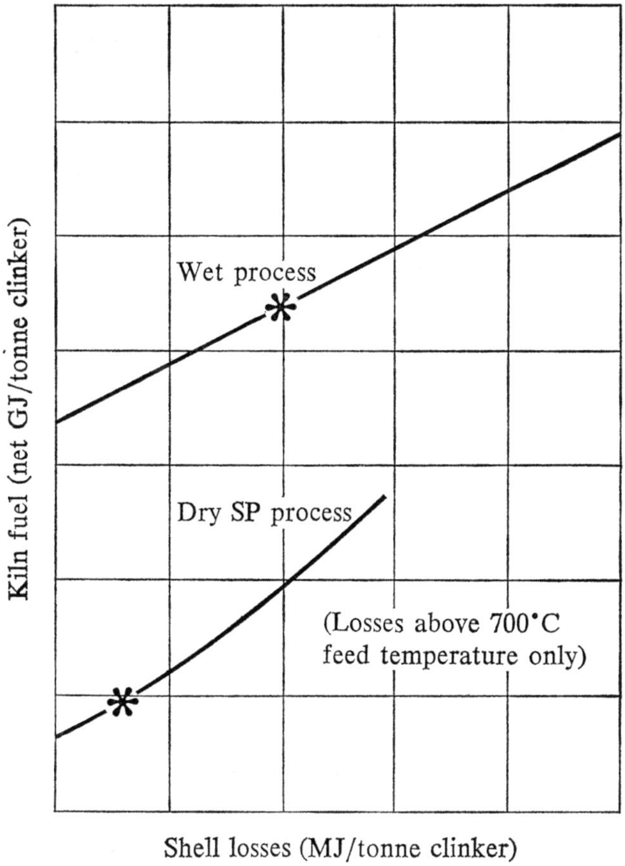

losses owing to reduced mass flow. Apparently, the poorer cooler performance counterbalances decreased kiln exhaust losses.

The shell losses of major importance on a dry process kiln are those occurring above a feed temperature of 700°C (i.e. the kiln and lower stage preheater losses) because these have most influence on the fuel required to obtain a certain minimum gas-feed temperature differential. In practice, heat losses below this temperature will represent some loss in preheater performance and kiln fuel. Figure 7·18 shows that a 40 MJ/tonne clinker increase in losses above 700°C leads to an increased fuel consumption of 50 MJ/tonne clinker.

Insulating refractories are an obvious way to reduce shell losses but they have had rather mixed success over the years. An insulating backing to basic bricks in the burning zone area does not seem practicable at present, especially on the larger kilns. On the other hand, modern hot face insulating bricks with thermal conductivities about one half normal firebricks are being successfully used. In Britain, the cost of these bricks is only a little more than normal firebricks and their life seems very similar. Generally, normal firebricks at the cooler end of the calcining zone are being gradually replaced with insulating types. Some experiments are in hand with high alumina insulating materials to extend the insulated region towards the burning zone.

Excess air and oxygen enrichment

Excess air and oxygen enrichment are considered together because they act in the same manner, although in opposite directions. Excess air increases gas mass flow and reduces temperatures for the same heat flow but oxygen enrichment reduces gas flows and increases temperatures.

Calculations show that for a wet process kiln, on the basis of a heat balance with constant kiln back-end temperature and clinker output, the use of air or oxygen enrichment would have very little direct effect on fuel consumption because of the compensating effect of secondary air flow rate on cooler recuperation efficiency. Reduction of excess air or the use of oxygen enrichment would have an advantageous effect for a dry process kiln because it increases the secondary air heat recuperation above the 700°C level.

These new heat balance calculations indicate that excess air and oxygen enrichment do not have a large direct effect on kiln fuel consumption. Nevertheless they can have a significant effect on output for a kiln limited by gas velocity and dust pick-up or by insufficient exhaust gas fan power. Excess air retards and oxygen enrichment improves output. As shell losses per kg of clinker tend to decrease with increasing output, and vice versa, it will be seen that excess air or oxygen can indirectly affect fuel consumption. However, cement manufacturers have never been able to make an economic case for using oxygen even when applying all factors in its favour.

Priorities for future development

The foregoing sections have indicated the effects of a number of kiln variables on fuel consumption, ignoring certain interactions between the variables for purposes of clarity. Cooler efficiency changes were an exception because changes in cooler air flow associated with the effect of changes in other variables could not be ignored. The question might rightly be asked as to what would happen to the fuel consumption if we were able to effect reasonable improvement to all the variables discussed and, in the final analysis, what would be the ultimate fuel consumption within the confines of the models used. These calculations have been made and the results are shown in Table 7·4.

The ideal and best practicable targets in the table set some interesting levels at which to aim, especially for the wet process plants. Some wet works in 1976 were in the region of 5·0 GJ/tonne which makes a very strong case for persisting with the better wet plants without conversion to the dry process until such time as an expanding market enables new dry process installations to be built. Rather than leaving the wet plants to run

Table 7·4
Conditions assumed for target calculations

	Dry process			Wet process	
	Ideal	Best practicable		Ideal	Best practicable
Slurry moisture (%)	25	25[1]	30[1]	–	–
Kiln dust loss (% of clinker)	0	3	3	0	0
Primary air (% total combustion air)	0	15	15	0	15
Inleak air (% total combustion air)	0	0	0	0	0
Excess air (% stoichiometric)	0	5	5	0	5
Shell loss (MJ/tonne clinker)	0	630	630	0	420
Back-end temperature (°C)	150	150	150	120	–
Cooler recuperative efficiency (%)	100	80	80	100	70
Calculated target fuel consumption (net GJ/tonne clinker)	3·34	4·36	4·77	2·19	2·87

[1] This limit may be dictated by slurry rheology.

down, priority should therefore be given to reducing slurry moisture, increasing dust return to 100% and reducing kiln exhaust temperature.

Priority for the dry process should be given to improving clinker cooler recuperation efficiency and reducing primary air and hood inleak air.

Electrical energy

The precise details of exactly what power is used by each item of plant on a cement works are rarely available but Table 7·5 shows the consumption figures broken down into different parts of the process for a number of plants.

There is considerable variation between works of the same type but this stems largely from differences in raw materials and hence in the energy needed for kiln feed preparation. This is particularly noticeable for the dry and semi-dry plants because they have harder raw materials.

Tables 7·6 and 7·7 show further detail for a typical dry process works. About 73% of the total power is used for raw meal preparation and cement

Table 7·5
Works' electric power consumption (typical works figures)

Process	Suspension preheater Hard materials	Lepol system Hard materials	Medium-hard materials	Wet process Soft materials (Works A)	Wet process Soft materials (Works B)
Kiln feed preparation (kWh/tonne clinker)					
Raw material getting and crushing	2·71	1·02	2·28	2·93	0·99
Raw meal milling/slurry preparation		35·77	29·60		
Blending and modulising	43·89	8·77	3·56	7·38	7·13
Sub total	46·60	45·56	35·44	10·31	8·12
Burning (kWh/tonne clinker)					
Firing and coal grinding	3·57	3·71	8·48	5·49	6·58
Burning-kiln-cooler-grate	22·83	17·05	14·99	22·46	22·52
Dust precipitators	4·48	0·26	5·35	3·64	5·56
Sub total	30·88	21·52	28·82	31·59	34·66
Total for clinker production (kWh/tonne clinker)	77·48	67·08	64·26	41·90	42·78
Total adjusted to cement basis (kWh/tonne cement)	74·03	63·60	60·16	38·79	39·79
Cement grinding etc. (kWh/tonne cement)					
Cement grinding	44·27	39·45	44·64	46·17	34·31
Filling, loading, etc	5·0	1·69	9·03	1·69	6·29
Total for cement manufacture (kWh/tonne cement)	123·30	104·74	113·83	86·65	80·39

grinding. Although most of this power is used to drive the mills, a sizeable proportion goes to power auxiliary equipment. For a typical cement mill section, about 70% of installed power is the mill drive but this consumes about 80% of the operating power. The raw meal preparation power will also include power for blending (about 3% of total power for this case). The burning process (including coal grinding and precipitators) takes about 20% of the total power used. The remaining 7% represents a large number of small items.

The importance of milling power is evident and must be considered in more detail.

Table 7·6
Average electric power consumption at dry works, 1975

	Electric power consumption (kWh/tonne)	(% total power) (Cement basis)
Kiln feed preparation (kWh/tonne clinker)		
Limestone quarrying	0·34	0·26
Limestone haulage, crushing, conveying	2·01	1·56
Shale quarrying	0·08	0·06
Shale haulage, crushing, conveying	0·28	0·22
Raw meal preparation	43·89	34·01
Burning (kWh/tonne clinker)		
Kiln fuel	0·27	0·21
Coal grinding	3·30	2·56
Coolers	6·33	4·91
Precipitators	4·48	3·47
Burning	16·50	12·79
Total for clinker (kWh/tonne clinker)	77·48	60·04
Total adjusted to cement basis (kWh/tonne cement)	74·03	60·04
Cement grinding etc (kWh/tonne cement)		
Cement grinding	44·27	35·89
Filling and loading	1·24	1·01
General	3·76	3·05
Total for cement manufacture (kWh/tonne cement)	123·30	100·00

Table 7·7
Cement mill of 2 250 kW installed power

Driven equipment	Installed power (kW)	(% total)
Mill drive	4 476	70·2
Separators	374	5·9
F.K. pumps and compressors	935	14·7
Miscellaneous fans and compressors	234	3·7
Conveying (mechanical)	173	2·7
Miscellaneous	175	2·8
Total	6 367	100·0

Raw meal crushing and milling

Comparison of ball mills and roller mills (verticle spindle mills)

Roller mills have been used for very many years for the preparation of raw meal and the comparison with ball mill circuits is one which is regularly under review. Early experience with roller mills in the late 1950s led to a critical awareness of the importance of skilled maintenance and the need for basically good materials and mechanical design. These features were, and still are, frequently regarded more closely than basic energy consumption. Nevertheless, roller mills can save energy for raw milling given suitable raw materials, and they are used successfully.

The commercially available roller mills differ slightly in principle. In one, the rollers are set at a clearance distance above a horizontal table and in others they are pressed on to a grooved track in the grinding table. Comparative performance data between the available types are insufficient to draw any firm conclusions as to whether one is more efficient than the others but as a group there is clear evidence that on mill power alone they require between 40% and 60% of the power of average ball mills. In addition, they will accept a somewhat coarser feed than ball mills and can thus make a saving on tertiary crusher power. Table 7·8 shows the differences.

Table 7·8
Comparison of typical energy figures for various milling systems (grinding to approximately 12% + 90μm)

Milling system	Tertiary crushing (kWh/tonne)	Mill only (kWh/tonne)	Fan and ancillaries (kWh/tonne)	**Total (kWh/tonne)**
Air swept ball mill	0·4	15·9	4·4	**20·7**
Double rotator ball mill	0·4	12·8	5·3	**18·5**
Short single chamber, bucket elevator ball mill	0·4	12·8	5·0	18·2
Vertical spindle/ roller mill	–	7·6	7·0	14·6

This difference almost certainly arises because the ground material is rapidly extracted from the grinding path and so minimises overgrinding, cushioning and re-agglomeration, which are inherent in virtually all ball mills where it is extremely difficult to extract ground powder from the mill charge even with considerable air sweeping. Much effort has gone into the design of ball milling circuits to try to improve matters. In general the air-swept mills have shown disappointingly high mill energy consumption and most successful systems have been based on either the double rotator principle (two short mills in series) or the short single chamber, bucket elevator, mechanical classifier mills with high circulating loads.

In making comparisons it is necessary to take into account total circuit power, not just the mills' alone, and some of the advantage of the roller mill is then offset by a rather high fan power which arises from the high pressure drop. This mainly occurs across the port ring to prevent coarse stone rejection and is an area for improvement. Taking into account the absence of tertiary crushing, the roller mill circuits overall show an energy saving over ball mills. It is difficult to get strictly comparative figures but Table 7·8 is based on plant and laboratory data to give a guide for a medium limestone. Obviously, factors such as moisture, grind fineness and the nature of the materials can influence the comparison. Approximately 6 kWh/tonne of clinker may be saved by using a roller mill system.

As indicated earlier, attention has to be focussed on mechanical factors particularly since there have been some disturbing failures, but the increasing costs of power present a case for looking afresh at roller mills where the conditions are favourable.

Finer stone pre-crushing before ball milling

Crushing is basically a more energy-efficient process than conventional ball milling because it operates in a size range where single particles are usually being dealt with and fine product is rapidly removed from the breakage zone.

In principle, it pays to crush until the number of particles involved causes cushioning and interference to approach the crowded situation in a ball mill. Then the simplicity of the ball mill dictates its choice for taking the process further.

The problem in practice is to decide at what size the crusher should finish and the mill take over. The nature of the material and the available ball size obviously influence the decision but the crucial factor is that the largest balls should quickly break the largest feed particles and here the top 2 or 3% of the feed grading matters more than the 80% passing size. If the top 2 or 3% of the grade cannot be crushed effectively in the time the material is in the zone of the large balls, it can rapidly build up in the mill or possibly in the circulating load. In either case drastic feed cut back is necessary. The subsequent grinding out of the mill causes overgrinding of the already fine material with considerable waste of energy.

All mills can suffer from these difficulties but the single chamber mills with classifying liners are vulnerable without the benefit of a diaphragm to hold back oversized material in the zone of the large balls.

It is difficult to generalise but given a knowledge of the material and the type of milling system it is possible to advise on the maximum lump size for efficient milling. The savings by crushing to an appropriate maximum lump size can be as high as 6 kWh/tonne (9 kWh/tonne clinker).

The tandem impactor/ball mill circuits are an extension of these concepts whereby an impactor tackles secondary crushed stone (up to 100 mm) immediately before a mill and, following classification, only the rejects (usually $\leqslant$10 mm) pass to the ball mill. With some hard limestones this finer crushing does not lead to any overall economy in power consumption but it may save around 13% in grinding power with suitable materials which yield many fines on impact or contain soft clay-like components.

In all these systems, considerations of wear and maintenance are crucial and may over-ride pure energy economy. Economies can be made by good crusher maintenance and ensuring that over-sized material is removed from mill feed.

Novel milling systems
There is always hope that one day an efficient grinding machine will be invented which is practicable and not susceptible to tremendous wear problems and high ancillary power.

Over the years, a number of novel machines have been tested but without success. The fundamental objective is to be able to tackle individual particles with the application of just sufficient loading to slowly break them. Various devices have been tried to achieve this, including rolls and ball races, but the problems of wear and the high cost of transporting material so far have defied inventors. Improved grinding efficiency may be found in the field of vertical spindle mills, with modifications to the table shape and the means of transporting material for classification.

More remotely, there might be some reward from carrying out comminution under abnormal conditions of temperature.

Cement grinding

Roller mills
As with raw material mills, the best prospect for substantially economising energy consumption in low surface area cement milling (other than favourably tailoring the clinker structure) will probably be the application of mills using the ring roll or ring ball principle. Experiences on raw material and coal grinding indicate that savings of up to 50% of mill power might be attained.

Wear problems are considerably greater for cement grinding than for raw material or coal grinding. In conventional ball mills, the metal of the media can be allowed to substantially wear away with little eventual scrap but the wearing parts of ring roll mills are large, fairly complex shapes from which only a relatively small proportion may be worn away before they have reached the end of their useful life. The availability of a wide range of wear resistant alloys and some recent, very favourable experiences with the track and ball life of a ring ball mill on hard coals indicate that progress might be made in this field.

Clinker feed size and pre-crushing
Clinker which is too coarse for the largest mill balls to crush effectively should not be fed to a cement mill, just as it should not be fed to a raw material mill. A steady feed, instead of the stop-start conditions due to overfilling the first chamber of the mills with coarse clinker, will give rise to economies, but too much pre-crushing is of no benefit. For example, if the top 5% of a clinker grading is between 25 and 50 mm and if it is scalped

off and crushed to less than 10 mm, the fineness of the mill feed would only increase from 8·3 to 8·7 m^2/kg which in the context of grinding to, say 300 m^2/kg is negligible. But, in removing a potential cause of hold up or blockages, such a move can be worth several kWh/tonne of power. Crushing further to a fineness level of less than 5 mm would only increase the feed surface to 9·8 m^2/kg which is hardly worthwhile. This is borne out by direct power measurements with finely crushed feeds for laboratory mills.

Overall, there is a case for some control of feed size to aid mill operation but not for a system of finely crushing all the clinker. The savings are difficult to quantify in general terms but, where oversize clinker is apparent, an analysis of the mill data should enable an assessment to be made.

Use of grinding aids

Grinding aids have been known for many years and broadly speaking they can be relied upon to save energy. As with all the factors discussed here, energy savings can be made in many ways although at some cost. In the final analysis, the question of whether the energy savings are economically worthwhile has to be faced and the decision whether or not to use a grinding aid is a clear example of this (cement quality standards permitting).

However, purely in terms of the potential for saving energy, full-scale experiments extending over six months with various grinding aids have shown savings from zero to 5 kWh/tonne for grinding ordinary cement and zero to 13 kWh/tonne for grinding rapid hardening cement.

Where grinding aids appear at first to be ineffective, control of the mill residence time by diaphragm adjustments might improve the results. There is also the possibility that some of the grinding aid effect is reduced by loss of the additive on to surface reactive clay associated with gypsum.

Grinding aids offer substantial power savings but the economic choice to use them or not depends on the relative costs of additive and power at any given time and on the value of any extra cement ground.

Open- and closed-circuit grinding

The trend over the past decade has undoubtedly been towards closed-circuit cement grinding and practically all large mills of 2 250 kW and above have been designed for closed-circuit operation.

The important claims for closed-circuit operation compared with open-circuit include better cooling, less likelihood of decomposed gypsum, better grit control and lower 45 μm residue (which helps 28-day strength), ability to grind to any chosen fineness without difficulty and better power consumption.

Without equivalent sizes of large open-circuit mill, it is difficult to make close comparisons but it seems reasonably clear that any energy savings from closed-circuit operation are marginal when grinding ordinary cement. This is not unexpected from small-scale test work but data from general plant and works tests over the period 1974–76 also demonstrate the point (see Table 7·9). Only the better open-circuit mills have been considered in the table.

Table 7·9
Power consumption of open- and closed-circuit mill systems

System	Power consumption (kWh/tonne) Mill only	Separators and elevators	Total power	Fineness (m^2/kg)
Open-circuit	31·1	–	31·1	320
Closed-circuit	28·7	2·2	30·9	321

For various reasons it is difficult to make similar comparisons from plant records for rapid hardening cement but in full-scale tests on closed-circuit plants the power savings again have not been as great as expected (typically 1 to 2 kWh/tonne in 55 kWh/tonne). However, the versatility of closed-circuit operation and the ease of control make it a necessary system where different and especially high surface area cements are needed but this duty might usefully be restricted to one mill.

At this stage the installation of closed-circuit grinding systems as potential power savers cannot be recommended except perhaps for improving an unsatisfactory old open-circuit system. However, a closer look is required at the systems currently installed to see whether they can be improved, especially for fine grinding.

Cement chemistry and grinding
The precise effects which the various constituents of clinker have on its grinding properties are not known but a certain amount can be inferred from studies of the grindability of various clinkers in relation to their constitution. These results relate directly to mill performance and energy requirements. There are two factors to consider—the clinker porosity and the intrinsic hardness of the solid material (crystals, glasses, etc.) from which the porous clinker pieces are built.

Porosity affects the crushing stage and the capacity of the larger mill balls to deal with the sizes of clinker fed. Clearly the more porous the clinker the less the milling problems, which can be very considerable with the very dense hard burned clinkers where no crushers are available. In general dense clinkers result from flux induration whilst porous clinkers are in the nature of a sinter with low fluxing.

The nature of the crystals and glasses influences the fine grinding process rather than the first chamber operation, and brittleness rather than hardness is the controlling factor. Brittleness depends on the relative proportions of alite (C_3S), belite (C_2S) and ferrite (C_4AF, etc). (In the cement industry the symbol C stands for CaO, S for SiO_2, A for Al_2O_3 and F for Fe_2O_3.)

In practice, it is rarely possible to fully control the raw mix to suit clinker grindability but, where the opportunity exists and with good steady kiln control, there is the possibility of saving grinding energy of the order of 10%.

Cement fineness and gypsum content
The importance of grinding only as fine as strictly necessary and with not less than the optimum level of gypsum has been stressed many times to plant operators, yet on some works appreciable variations still occur. An improvement here offers significant energy savings.

Table 7·10 and 7·11 show, in a typical case, the extra power consumption and loss of output for a 2 240-kW mill of various deviations from fineness and gypsum targets.

Table 7·10
Effect of overgrinding

Fineness (m^2/kg above target)	Power consumption (kWh/tonne)	Extra power consumption (kWh/tonne above target)	2 250-kW mill output (tonne/h)	Loss in output (tonne/h)
Ordinary Portland cement				
Target	34·0	–	65·6	–
10	35·5	1·5	62·8	2·8
20	37·0	3·0	60·3	5·3
30	38·9	4·9	57·3	8·3
Rapid hardening Portland cement				
Target	60·3	–	37·0	–
10	63·8	3·5	35·0	2·0
20	67·0	6·7	33·3	3·7
30	71·2	10·9	31·3	5·7

Table 7·11
Effect of grinding with below target gypsum levels

Gypsum level (% of clinker)	Power consumption (kWh/tonne)	Extra power consumption (kWh/tonne)	2 250-kW mill output (tonne/h)	Loss in output (tonne/h)
2·6 (i.e. target)	34·0	–	65·6	–
2·5	34·5	0·5	64·6	1·0
2·4	35·1	1·1	63·5	2·1
2·3	35·7	1·7	62·5	3·0

Other uses of electricity
Other than the mill systems already discussed and coal grinding, for which many of the comments on raw milling also apply, the remaining power is consumed by a few main items and a large number of small units of comparatively low consumption. Economies in this field fall mainly into the area of good maintenance and general housekeeping.

Fans

Fans can be very wasteful in power if not used with care. Massive inleaks of air can occur undetected and checks of the oxygen levels at kiln and precipitator fans are made as a general economy measure. Table 7·12 shows the effect on fan power of various oxygen levels at a given fan suction. With kiln exit gas oxygen levels of 1 to 2%, it is common to find levels above 7% at the inlets to chimneys.

Table 7·12
The effect of air inleak on fan power

Oxygen (%)	Relative power
1	1·0
3	1·074
5	1·182
7	1·324
9	1·485

Blending

The power consumed in blending depends more on the total blending capacity than the throughput. Typical raw meal fluidised blending silos require around 4 kWh/tonne clinker, whereas a blending stockpile uses 0·4 kWh/tonne clinker. Because of this, a number of research studies are in progress to assess the merits of basing the blending on one or the other or on some combination of both. Wet process blending by air uses about 0·5 kWh/tonne clinker.

In either the wet or dry process it cannot be emphasised too strongly that it is false economy to save power at the expense of blending quality. This can result in increased fuel consumption and increased cement grinding costs which far outweigh any small savings from cutting back on good blending technique.

Conveying

Pneumatic conveying uses 10 to 100 times more power than belt conveying. In practice these methods are not often interchangeable and it is usually factors other than power consumption which determine the method to be used.

Priorities for future development

Useful economies in power are possible by attention to the various points already discussed but the area where the greatest impact needs to be made is in milling and a great deal could be achieved by making sure that mills are fed steadily with a feed size they can handle. The greatest potential for economy is in the development of ring roll and ring ball mills for grinding raw meal and cement.

Auxiliary fuel

In terms of the primary heat and power requirements for cement making the auxiliary fuels used for road transport and space heating represent only a small percentage of the whole. In absolute terms, however, the gross cost for a large cement making operation is considerable. For example the cost of diesel oil in the UK in 1975 was over $4 million.

Diesel fuel

Economies on the transport side depend very much on local maintenance of the lorries and good distribution planning. Typically, fuel consumption for road transport is of the order of 39 tonne km/litre. This compares with rail performance of about 283 tonne km/litre.

Lepol system—raw material drying

With regard to general heating, the oil used in Lepol plants for raw material drying amounts on average to around 300 MJ/tonne clinker. This is an area for economy by using spare cooler heat or cheaper fuels.

Office and other general heating

There is some advantage in using kiln system waste heat to provide heating for adjacent workshops and offices. There are various sources of such heat, for example in grate cooler exhausts, in kiln exit gases and in radiation from the kiln shell. A number of experimental projects are in progress, ranging from a project to sell heat for domestic heating (using a heat exchanger in kiln gases) to relatively small banks of tubes arranged to pick up kiln shell radiation to provide hot water for the workmen's showers.

Alternative fuels to oil and gas

The rapid rise in oil prices and the general shortage of gas have caused many cement makers to search for cheaper alternative fuels. While not strictly economising on overall heat requirements, this can lead to cost economy and with some materials can be environmentally desirable by using potential fuels which would otherwise be wasted.

Coal

Coals, unlike oil or natural gas, contain considerable quantities of ash and water. The ash and moisture contents of coals from different sources can vary widely, resulting in a wide spread of calorific values. The calorific value of dry mineral-free coal also varies somewhat with rank but this is unimportant compared with the effects of ash and moisture. Unlike coals, heavy fuel oils obtained from different sources have very similar calorific values. Natural gas from different sources varies considerably in composition but the range of calorific values is less than with coal. Further, oil or gas from a given source will remain substantially constant in calorific value but coal from the same colliery may fluctuate significantly in ash and moisture content and hence calorific value.

Burning coal instead of oil or gas has a major effect on the chemistry of the cement making process since the composition of kiln feed has to be adjusted to accommodate absorption of ash. This arises because the ash causes a significant lowering of the lime saturation factor (LSF) necessitating a kiln feed target LSF considerably higher than that required in the resultant clinker. As a rough guide, the absorption by the raw feed of a typical ash with a silica content in the region of 50% in the proportion of 1% ash to clinker lowers the LSF by about 4% although the effect will be greater for an ash with a significantly higher silica content.

The extent of ash absorption is largely governed by the type of process. It will be apparent that careful blending of the coal is highly desirable to ensure stable kiln chemistry, especially where ash contents are high.

A second important aspect is that the ash has to be effectively absorbed into the clinker if the clinker quality is to be maintained. Particularly with high ash coals, the failure to absorb ash effectively can cause a heterogeneous clinker and necessitate hard burning. The problem can be contained by fine grinding of the coal and by suitable choice of fuel injection conditions. Provided the milling/firing system is matched to the coal fuel and kiln process, coals of up to 35% ash can be used.

Other compositional differences between coal, oil and gas such as the proportions of sulphur and alkalis have little effect on raw material composition or product quality. Carbon and hydrogen differences have no effect. Sulphur, alkalis and chlorides in the fuel can have an effect on kiln operation while the composition of the combustibles influences flame properties and quantities of combustion gas. In preheater dry process kilns, the preheater acts as a filter to remove the sulphates in the gas stream and as a scrubber in which sulphur dioxide combines with lime or alkaline oxides. Hence most of these sulphates are recycled to hotter zones of the kiln, causing hard-bound deposits and inefficient operation in certain circumstances.

Minestone

Minestone is a low grade fuel commonly screened out and rejected at collieries. The calorific value varies but is typically around 5 GJ/tonne, compared with coal at around 30 GJ/tonne. Minestone is a mixture of coal and shale and provided it is blended to achieve a known chemical consistency it is possible to replace part of the argillaceous component of the slurry by the shale in the minestone.

The minestone can be introduced to a wet kiln by insufflation as a dry powder close to the normal firing pipe or alternatively it may be ground in with the slurry. In either case a kiln would not operate entirely on minestone because of the quantity of minestone involved. It offers a part replacement of fuel at a relatively low cost if the collieries are near and seek only a nominal selling price.

A further advantage is that by reducing the shale or clay proportion in a slurry it is often possible to operate with 1 or 2% lower moisture and gain further advantages to fuel economy.

For the dry process, the technique of firing a low grade fuel into a special furnace between the kiln and preheater has already been used at a plant in Dotternhausen, Germany, and this system is commercially available. Various major plant suppliers are interested in the same equipment and offer slightly different designs.

Whatever technique is eventually adopted, the use of minestone is clearly a possible method of economising on fuel cost provided adequate chemical control can be achieved.

Domestic refuse

Pulverised refuse from which certain heavy components have been removed is also an alternative fuel. It has a calorific value of around 12 GJ/tonne. Experiments have already been successfully completed showing that the ash from the combustion of the refuse can be absorbed into the clinker without any harmful effects. Techniques for conveying and firing the refuse have been developed and two UK wet process works receive local refuse to supplement kiln fuel. In general, the addition rate is of the order of a 15% replacement of kiln fuel.

Waste oils

Cheap waste oils are sometimes available and arrangements have been made to fire such oils as a partial replacement for the normal kiln fuels.

Low flashpoint oils are generally best avoided because of fire hazards and the filtering and pumping systems must be adequate to deal with considerable solid ash and water contents.

Utilisation of low grade heat and other inter-industry possibilities

Hot exhaust air from the clinker cooler might be used for drying agricultural crops, for example. In some dry process plants there is more hot kiln exhaust gas than is needed for raw material milling-drying and the excess might be used for raising low pressure steam or hot water to be used inside or outside the works.

Other possibilities for inter-industry co-operation have been mentioned previously including the burning of refuse and other low-grade fuels to economise on the prime kiln fuel and the use of colliery rejects as a cement raw material. Additionally, lime-containing slags might be used as a raw material and potash-rich cement kiln dusts could be used in fertilisers.

University research

University research projects could be chosen from the following list.

Kiln process

- Design of efficient and compact cyclones for gas-powder heat exchange.
- Prediction and control of ash deposition from the flames of high ash fuels.
- Design of pulverised coal firing pipe injectors to reduce injection velocity, fanpower and fan wear.
- Improved refractories (better life, lower thermal conductivity, lower cost).

- Uses for large quantities of low grade heat.
- Studies of gas and solids flow and heat transfer in furnaces.

Comminution

- Design of more efficient machines capable of scale-up to 100 tonne/hour or more, with acceptable repairs costs when grinding abrasive feeds.
- Better understanding of grinding characteristics of materials.
- Better understanding of the mechanism of grinding aids.

Chemistry

- A chemical approach to lower energy requirements, particularly energy for cement clinker grinding.

Rheology

- Additives to maintain the flow of cement feed slurries with low moisture content, particularly in the presence of chloride and sulphate ions.
- Design basis for a truly continuous pressure filter.

8 Paper and board

Chapter contents

Introduction

Energy usage

Estimates of the energy intensity of some products of the paper industry have been obtained as part of the Department of Industry Energy Audit Scheme (Palmer, 1977) and are quoted in Table 8·1, together with estimates of the output for each product in 1976. The energy requirement in each case is broken down into three elements: energy input to the pulp, primary fuel energy for other processes and electrical energy for other processes (given as primary fuel equivalent). As most of the pulp used in the industry is imported, the energy requirement of this input should in principle be excluded from the UK figures. The energy intensities range from approximately 14 GJ/tonne for fluting (the filler section in corrugated board, made from 100% waste fibre) to approximately 38 GJ/tonne (29 GJ/tonne excluding input pulp energy) for soft tissues where the high speed of production required because of the rapid drying rate of the product results in a particularly high electricity consumption. The sectors with the largest output (printing/writing grades of paper and board, 24 and 21% of the total output respectively) are also amongst the most energy intensive.

Table 8·1
Output and energy requirements of some products of the UK paper industry

Product	1976 Production[1] (tonne×10³)	Energy inputs (GJ/tonne) Pulp manufacture	Primary fuel process	Electrical[2] process	Total
Fluting (100% secondary fibre)	420	0·2	10·0	3·8	14·0
Liner grades	240	2·4	7·0	12·6	22·0
Case liner grades	[3]	4·0	10·5	8·5	23·0
Board	880	2·5	16·0	8·5	27·0
Coated paper grades	[3]	10·5	13·5	6·0	30·0
Printing/writing paper	980	11·0	18·5	7·0	36·5
Soft tissues	360	9·0	13·0	16·0	38·0

[1] Total 1976 production, 4 100×10³ tonne.
[2] Quoted as primary equivalent.
[3] Figures not available.

The trends in the energy intensity of the paper industry as a whole for the period 1955 to 1975 are illustrated in Table 8·2 (Abott, 1977) and the changes in sources of energy purchased over this period are shown in Table 8·3. The fossil fuels purchased are mainly consumed in steam generation used for the purpose of in-house electricity generation and subsequently to satisfy process heat requirements.

Table 8·2
Energy cost of paper[1]

Year	Production (tonne×10⁶)	Energy input			
		Total delivered		Primary fuel equivalent	
		(GJ×10⁶)	(GJ/tonne)	(GJ×10⁶)	(GJ/tonne)
1955	3·35	111	33·3	125	37·4
1960	4·14	111	26·9	133	32·1
1965	4·61	121	26·3	148	32·1
1970	4·94	124	25·2	151	30·6
1971	4·37	115	26·3	142	32·4
1972	4·37	114	26·0	142	32·4
1973	4·71	112	23·7	141	30·0
1974	4·57	104	22·7	131	28·8
1975	3·66	93	25·5	120	32·9

[1] Sources: Census of production, Annual abstract of statistics and Digest of UK energy statistics.

Table 8·3
Forms of energy purchased by UK paper industry[1]

Year	Form of energy (% of total)			
	Coal or coke	Oil	Gas	Electricity
1955	93·4	4·9	0·1	1·6
1960	81·2	15·3	0·1	3·3
1965	68·4	27·6	0·1	3·9
1970	55·7	39·7	0·1	4·5
1971	43·3	46·4	5·6	4·7
1972	33·7	46·6	15·2	4·6
1973	31·3	45·9	17·7	5·1
1974	28·3	44·8	21·6	5·3
1975	28·1	43·9	21·8	6·2

[1] Source as Table 2.

A schematic illustration of the main energy flows in the UK paper industry is given in Figure 8·1.

A breakdown of the costs of manufacturing various products within the paper industry in 1971 is given in Table 8·4 (Webzell, 1972) and shows that materials and labour costs were the most important items. Recent information indicates that although the oil price rises of 1973/4 increased the energy cost element to 12 to 14% of total costs, subsequent rises in labour and raw material charges have almost returned the costs to the pre-1973 ratios (Newton, 1977).

Figure 8·1
Main energy flows in the UK paper industry, 1973

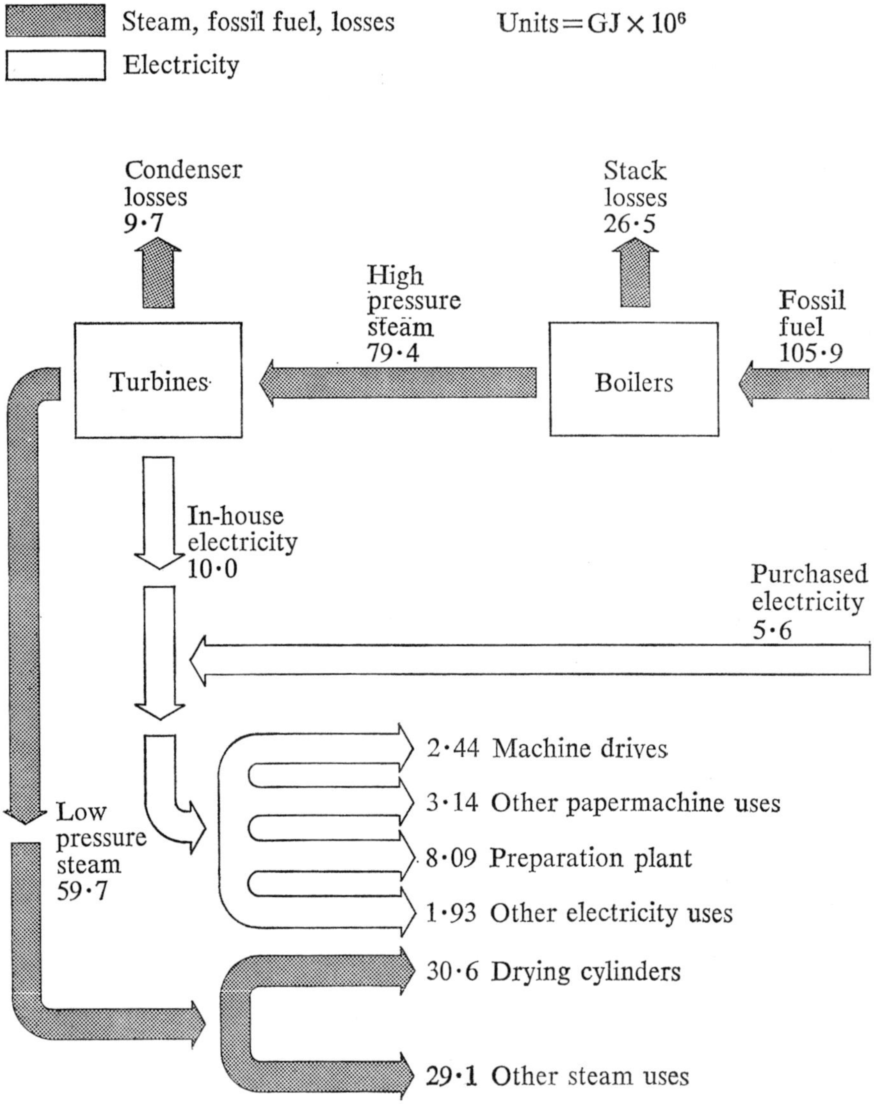

Table 8·4
Breakdown of costs within the paper industry, 1971 (Webzell, 1972)

Breakdown	Share of total cost (% of total) Newsprint	Packaging paper	Hardboard
Labour	12	17	29
Materials	70	64	50
Energy	6	7	9
Maintenance, etc	6	6	8
Others	6	6	4
Total	100	100	100

Process details

The generic term paper covers the following categories of products:
- board—multi-ply with basis weight 220 g/m^2 or greater
- paperboard—multi-ply paper with thickness at least 0·25 mm
- paper—single-ply, including tissues, etc.

While the processes involved in the preparation of the stock (the fibre-water mixture) are similar for all types of paper, the techniques employed in sheet production do vary, as explained in the following sections.

Wood pulp manufacture

Since the majority of the wood pulp used in the UK is imported from timber mills in the form of dry pulp, its manufacture is not relevant to this study. However, the methods used do have some bearing on the stock preparation process. The methods are listed below.
- Chemical processes in which wood chips are chemically treated in pressurised steam-heated digestors to remove the lignin from the wood and allow the separation of the cellulose fibres. Used in the production of high quality pulps.
- Mechanical processes where the wood chips are ground to produce a pulp of lower quality.
- Semi-mechanical processes in which the wood chips are sometimes pre-heated and/or chemically treated before mechanical grinding to produce a pulp which has superior properties to normal mechanically processed pulps but has not incurred the expense of full chemical processing.

Stock preparation

Primary fibres

The object of the stock preparation process is to separate the fibres in the pulp and subsequently modify their form by mechanical treatment, so increasing their area by fibrillation to improve the bonding in the finished product. By creating conditions that assist in the free uptake of water into the fibres, the fibres are made more flexible and will easily deform under the surface tension developed during paper drying and so develop bonding between the fibres in the web.

The exact sequence of events in the stock preparation is dependent on both the type of pulp used and the type of paper to be produced. Firstly, imported pulps have to be re-suspended in water and their fibres separated. This process is carried out in slushing units (which are described below). The extent of the refining treatment given to pulps in their preparation varies depending on whether they are of mechanical or chemical origin. With mechanically processed pulps, the lignin still associated with the cellulose fibres will inhibit the process of fibrillation so restricting the value of any refining carried out, but a considerable improvement in the bonding characteristics of chemically processed pulp can be achieved through fibrillation accomplished in refining.

Slushing units

The breakdown of the pulp into separate fibres is usually carried out in a form of centrifugal disintegrator, such as the hydrapulper. This consists of a vertical bowl-shaped tank (varying in diameter from 1 to 6 m with an average depth of 3 m). The action of a multi-vane rotor in the base creates a vortex in the pulp which ensures good mixing and promotes the separation of the fibres in the pulp. The equipment operates with a stock consistency of 6 to 12% and the energy required in its operation is approximately 70 MJ/tonne for continuous operation and 140 MJ/tonne in batch running.

Refining

The mechanical modification of the separated fibres in the stock is accomplished in refiners of two basic designs, conical and disc.

A conical refiner consists of a cone on the surface of which are bars that rotate inside a casing about a horizontal axis. The stock, which is fed in at the apex, is refined as it passes between the bars and the casing, the severity of the action being controlled by the spacing between these two members. A very approximate figure for the energy consumed in this type of refiner is between 800 and 1 600 MJ/tonne, but this is strongly dependent on the fibres treated, the size and mode of operation of the unit and other variables.

Disc refiners have been widely adopted in recent years as part of a drive to improve the efficiency of the refining operation by reducing specific power consumption and increasing control over the process. Disc refiners consume less energy than conical refiners (about 20% less) for a given fibre treatment. Typical energy consumption is 500 to 700 MJ/tonne.

The stock is fed to a disc refiner through an opening in a stationary disc into the gap between this and an opposing rotating disc. The fibres gradually move towards the edges of the discs and the refining action is accomplished as they pass between grooves and bars on the contacting surfaces. Close control of the gap between the discs is required for efficient operation—too large a gap will permit large quantities of fibres to pass through the working area untreated. The relative speed of rotation of the disc plates has an influence on the type of refining action obtained. Slow speeds (less than

1 200 m/min) produce a predominantly cutting action with little fibrillation while high speeds (greater than 1 800 m/min) give mainly fibrillation with little cutting. Generally a combination of the two processes is desired and typical speeds of operation are around 1 500 m/min. Further variables in the operation of disc refiners are the angle of the bars to the radius of the disc face (a trailing bar angle tending to throw the stock from the discs while a leading bar angle tends to hold back the stock and prolong treatment) and the depth and height of the grooves and bars respectively.

Other processes involved in stock preparation

Of the remaining processes in stock preparation, the main energy demand occurs in pumping the stock between process stages and in keeping it circulating in holding tanks.

Energy and materials aspects

Refining is regarded by some as an inefficient process with efficiencies of the order of 0·1% of the theoretical energy requirements. The assessment of the condition of the stock often relies largely on operator judgement, although some automatic devices designed to monitor the condition of the stock during processing have been developed. A lack of control over some systems has meant that it has been difficult to monitor their performance, and thus prevent an undue proportion of the energy supplied being converted to an increase in stock temperature instead of a mechanical modification of the fibres.

The high electrical energy requirements of this process and the low theoretical efficiency of present techniques suggests that further study of refining processes, possibly involving alternative approaches such as ultrasonics, is justified.

Some opportunity for material conservation exists by increasing the filler content of paper in an effort to conserve wood resources and reduce material costs. Fillers (white mineral particles that fill in spaces between fibres, help in printing, etc.), sizes (starch and resin treatments that render the paper resistant to penetration by water) and dyestuffs can all be added before the stock enters the papermaking machine.

The substitution of other fibrous materials for wood pulp such as straw pulp and willow grown as a crop, is also being investigated by the industry to reduce material costs.

Secondary fibres

The fibres in recycled waste paper have previously undergone a refining process and all that is required in their preparation for re-use is separation from the sheet and removal of any impurities such as coatings, printing ink, etc.

The separation of the fibres begins in a hydrapulper from which the stock is transferred to a de-flaker, where the final process of individual fibre separation is accomplished with a minimum of further damage to the already mechanically treated fibres. For some applications, such as the filler ply in board products, the secondary fibre pulp can be used without any further treatment other than washing and the removal of foreign bodies.

Before use in high quality applications, the pulp requires chemical treatment in the form of de-inking, asphalt removal, etc.

The first operation in the de-inking process is usually carried out in conjunction with the slushing stage when chemicals are added which are able to separate the ink particles from the fibres when the stock is heated to 60 to 80°C. Heating to this temperature and maintaining it as the chemical action occurs consumes a significant amount of energy. The development of new, lower temperature processes could produce worthwhile savings, as could the development and adoption of inks which are easier to remove. The removal of the ink particles takes place in flotation cells and is accompanied by the loss of 5 to 10% of the recycled material. After flotation, the stock is concentrated ready for the papermaking process.

Some recycled paper contains asphalt or other waterproofing materials which have to be treated to avoid production difficulties from deposits on machines and black specks in the finished product. The total removal of the asphalt from the stock would be a difficult task and so instead it is dispersed by raising the temperature of the stock above the asphalt melting point (approximately 140°C). Steam consumption in heating the stock to this temperature is high and, in order to minimise the energy required, the stock is usually thickened before treatment.

Other types of impurity in secondary fibres can cause problems, in particular stickies, which are the residue of self-adhesive finishes, that can result in production problems through roll fouling and product damage.

The removal of impurities from secondary fibres could provide some opportunities for research and development aimed at energy savings and extending secondary fibre applications.

Paper-making

The stages involved in the production of a wide range of paper goods may be considered under three main headings: wet end operation, mechanical water removal and drying operations.

Wet end operation

The basic systems

Most single-ply paper products are produced on Fourdrinier machines in which the web is formed by the projection of the stock from a head box (a form of metering device) onto a continuous travelling wire in the form of a mesh belt. At this stage the water/fibre ratio of the stock is about 200 but is reduced to about 2 by mechanical water extraction before the thermal drying techniques are employed. Water drains from the web through the wire, which is often moved from side to side in order to assist the water loss and to help intermesh the fibres. Towards the end of its travel the wire passes over suction boxes which remove further water.

Multi-ply products require the combination of several independent webs to form the finished board and, although combination machines made up of several Fourdrinier wire sections have been used in board production, the most common technique involves the use of cylinder moulds.

A cylinder mould consists of a wire covered cylindrical drum which rotates in a vat containing the stock. Individual fibres from the stock are picked up by the wire surface and contribute to the development of a uniform web. The water suspending the fibres drains through the cylinder and is discharged from the system. The level of the stock in the vat is usually three-quarters of the height of the cylinder and it flows continuously with a uniform consistency across the machine.

Two types of basic cylinder mould machines exist: uniflow with the stock and cylinder moving in the same direction (which gives a uniform sheet but low capacity) and contraflow where the stock and cylinder move in opposite directions and the sheets are produced at a faster rate but to a lower uniformity. The wet mat of fibres produced on the cylinder mould surface is either transferred to a travelling felt by a couch roll in the case of single-ply products or is carried by a continuous wire belt to subsequent cylinder mould vats where successive layers of web are added until the desired thickness is achieved.

Improvements

Many variants of the two basic wet end processes exist, but only those developments with particular relevance to energy or materials use are described here.

Most effort in single-ply product systems appears to have centred on developing improved methods of mechanical water removal, discussed later, or the secondary fibre chemical treatments already described. In board and paperboard technology, however, particular attention has been given to improving methods of sheet formation in order to improve performance through increased production, to developing the use of lower grade secondary fibres (particularly for middle plies) and to using new raw materials.

A problem in the design of conventional wet end systems used in the production of multi-ply products is that once the first web has been transferred to the top side of the wire connecting the individual cylinder moulds, drainage of subsequent webs through this consolidated layer becomes progressively more difficult. An increase in the production speed can apparently only be achieved in these circumstances by increasing the length of the wire section and therefore the size of the wet end part of the machine. An alternative solution to this problem is to obtain drainage from both sides of the web, and two devices which employ this technique are the Inverform and Verti-forma machines (although the latter is used mainly for single-ply products).

In the Inverform system, stock is projected from a head box between two wires which then pass through the contact between a large cellular roll and a pressure roll. At the exit from the roller, an autoslice (basically a scraper which rests on the surface of the top wire) collects water from the upper surface. Further removal of water from this surface is achieved by suction boxes. The technique is reported to allow at least a doubling of production rates over normal cylinder mould units and produces a high quality, more uniform product.

The Verti-forma is also a twin wire device but operates in a vertical plane with a head box and metering roll controlling the release of stock between a pair of converging wires which carry the web through a series of suction boxes to extract water from both sides of the web. The product has more uniform cross-section and equal through-thickness properties.

The dry vat cylinder mould is a development of the basic cylinder moulds in which the stock only comes into contact with the cylinder for about a quarter of its circumference. The stock is dispersed into the mould surface from a conforming head box and the associated water drains through the cylinder leaving the web of fibres on the mould surface. The web is then transferred to a travelling felt which links the units in the normal way. Dry vats are reported to produce sheet of more even weight as a result of a constant head across the machine. There is also a substantial reduction in the number of fibres washed off the mould during the process of web development compared to conventional vats with long passages through the stock. In addition, the elimination of uncontrolled stock flows improves the appearance and profile of the sheets produced.

Details of other cylinder mould units can be found in standard references (e.g. Higham, 1970).

Energy savings

The stock consistency used in paper-making is largely determined by the level necessary for satisfactory sheet formation with the head boxes available. Although an increase in stock consistency would be of little interest in single-ply products where natural drainage freely operates to reduce the water/fibre ratio from 200 at the head box to less than ten at the presses, the energy use involved in multi-ply products, where free drainage of surplus water is often restricted, could benefit from the use of increased consistency stock through improved head box design. Savings, however, would have to be offset against any increases in power requirements which might arise in having to pump the denser stock.

Mechanical water removal

At the end of the forward travel of the wire in the wet end stage, the web is transferred to travelling felts which carry it through the press section.

Methods

In the simplest form of press, the plain press, the web and felt pass through the contact between a pair of rolls and are separated at the exit in an effort to prevent the water, expelled from the web into the felt, re-wetting the web. An increase in the amount of water removed in the press may be achieved through greater sheet compression although the pressure that can be applied is limited by the risk of deforming the wet sheet. In practice, the progressive removal of water is achieved by the gradual application of increased pressure as the web passes through a battery of presses.

These limitations are to some extent overcome in the fabric press which includes a third layer (in addition to the web and felt) in the contact between the press rolls to act as a non-compressible space for the free passage of water released from the web and the felt. The presence of this third layer apparently allows the use of higher pressures without the risk of crushing the sheet and as a result has the following advantages:

- production of drier sheet at higher speeds than is possible in plain press operation;
- opportunity to either increase production rates or to reduce steam consumption during the drying process;
- improved retention of fillers through a more uniform moisture profile across the sheet;
- smoother sheet surface because of the higher pressures applied.

The felts in the press section are dried either by suction boxes or an air knife. Other developments in press design include the use of suction presses and air drying for single-ply products, which is considered later.

Energy savings

At the point of entry into a press section, the moisture content of the web is typically 80% and this is reduced during pressing to approximately 64%. Performance of press sections varies considerably between establishments, one estimate of exit moisture content is between 62 and 69% and some installations are able to operate with a moisture content at entry to the dryer section of 58 to 60%. One US mill is even reported to operate with a sheet moisture content of 53% at the end of pressing. Considerable energy savings in the dryer section result from improvements in mechanical water removal. For example, a reduction in mean moisture content from 68 to 64% at entry to the steam dryers could result in fuel savings of about 1 500 MJ/tonne, allowing for both boiler efficiency and condensate return.

While most developments in mechanical drying technology have arisen from a need to increase output from an established press section, the progress achieved in recent years has also provided opportunities for direct reduction in energy consumption in paper drying. It has been stated that the opportunities for improvements in water removal by increased pressure have now been exhausted. Further savings would arise through improvements in felt performance and by modification to the mechanical drying process by, for example, exploiting the fall in water viscosity with rising temperature to improve press efficiency.

Dryer section

Steam-heated cylinders

The remaining water in the sheet is removed by passing it on a felt around a series of 20 to 50 steam-heated rotating cylinders. The final dried paper has a moisture content of 4 to 7%. Several variations in the design of this stage exist. In the MG machine, for example, the sheet is carried on a special felt between a heavy rubber-covered roll and a large, highly polished steam-heated cylinder. The web sticks to the glazed roll and is removed nearly dry at about two-thirds of the way round, having received a machine glazed finish, which reduces the need for calendering in the finishing stages.

The detailed mechanism of water removal by evaporation from paper is complex. Drying proceeds at a constant rate while there is water at the surface of the paper but, in the later stages, the migration of water from the moist interior to the dry surface becomes the rate determining process. The energy required in the evaporative drying of newspaper is thought to lie in the range 1·2 to 2·0 kg steam/kg water evaporated. Accepting the figure of 1·4 kg/kg as representative of good practice then it is possible to calculate that conventional steam-heated cylinders operate at an efficiency of about 64%, or 45% if boiler efficiency is included.

The incentive for improving drying techniques develops from the need to increase overall productivity of a papermaking plant since the drying stage at present dictates the speed of the overall process operation. While the basic efficiency of the steam-drying cylinder appears to offer limited scope for improvement, several possibilities exist for improvements in paper drying techniques which will improve productivity and at the same time provide some energy savings. The problems encountered in using conventional drying cylinders include:

- high thermal inertia of steam-heated cylinders which cannot respond quickly to required changes in drying rate;
- difficulty in maintaining a uniform moisture profile across the machine;
- difficulties in achieving good control of drying by successful monitoring of inlet and outlet moisture profiles and by measuring moisture levels within the dryer;
- difficulty in adequately removing moisture-laden air from the vicinity of the product being dried;
- low efficiency of water removal in the later stages of drying.

Several measures have been considered as possible solutions to these problems and are briefly described below.

Radiant heat drying

The technique of radiant heating has the potential for extremely high heat transfer rates and is thought to be capable of drying with an efficiency of 50 to 60%, although it is less efficient with dryer materials. Development of this technique has been limited by the higher cost of its fuels (gas and electricity) in comparison with steam. Its use is consequently limited to certain special functions such as quick drying of coatings immediately after they are applied to paper, which improves both quality and speed of production. It has also been used to increase the temperature of the web during mechanical pressing to aid water removal and to pre-heat the web entering steam dryers. Some advantage has been reported for drawing the hot combustion gases produced by a radiant heating system through the product with the aid of suction rolls.

Dielectric heating

The wide-scale application of dielectric heating would be uneconomic but it has certain inherent features which could make its application attractive. The main advantage is the ability to produce uniform moisture profiles, since most heat is generated by this process in the wettest areas. In addition, the system has other advantages.

- Response times are quick.
- Energy loss at idling is zero. The system can be switched off and restarted within 30 seconds. This feature could be attractive to the paper industry where no-load energy requirements are thought to be as high as 50% of the total energy consumed in some large mills.
- Control of energy requirements is automatic as the system only draws the energy required by the moisture content of the product concerned. This technique would appear to be particularly suitable for improving drying efficiencies at the dry end of the process where water removal is limited by its migration to the paper surface.

Air drying

A major problem in achieving effective drying is not raising the temperature of the material to be dried but removing from the products the water already evaporated and contained in the surrounding air. If the relative humidity of the air in the vicinity of the web is allowed to rise too high, further evaporation is retarded and production therefore slowed. Several systems incorporating air drying in one form or another exist.

- *Open and closed hoods.* Air flows in drying sections are most easily controlled by open and closed hoods which take the form of canopies over the machine to trap and exhaust moist air. Closed hoods which enclose most of the paper machine are more effective in reducing steam demand in dryers and exhaust smaller volumes of air compared with open hoods. The higher temperature of the exhaust from closed hoods offers greater opportunity for waste heat recovery. Current estimates suggest that 50% of Fourdrinier machines have no hood and only those installed in the last ten years are fitted with closed hoods.
- *Pocket ventilation.* Because of their enclosed position, the spaces between cylinders (the machine pockets) tend to trap moist air. To combat this, air heated by steam coils is introduced into the pockets. The energy requirements of these systems are small in relation to the 10 to 15% increase in production their installation can bring about.
- *High velocity hoods.* A device to blow air at high velocity onto the paper surface to break up the surface layer of saturated air improves the rate of evaporation. Such systems can significantly increase production from a machine and also reduce the energy required in drying. Several attempts

to use ducted hot air in conjunction with high velocity hoods have been made, but suffer from producing excessive damage to the web, even though it is contained between wires.

- *Through-air drying (single-ply products).* Conventional air systems aim to remove saturated air from the dryer unit and so speed up evaporation from the paper surface. As with steam dryers, these systems become less efficient in the later stages of drying, but this is overcome in through-air drying processes. In this approach, hot air passes through the web (picking up moisture during its passage) and is exhausted by passing through the perforations in the surface of the cylinder which is under vacuum. The hot air is usually supplied by a high velocity hood. The efficiency of such systems is reported to be in excess of 70% and, by overcoming the internal diffusion problem, allows high production rates to be achieved.

Alternative paper-making processes

Dry paper forming

The development of a commercial dry paper forming process could result in energy savings amounting to 50% of the energy used in conventional plant. Briefly, the following stages are involved in one example of this technique of paper forming:

- dry pulp is dispensed onto a mesh belt which moves over vacuum boxes, to form a thin uniform layer;
- the web is dampened to assist bonding and some chemicals may be added;
- the web then passes around a large cylinder for drying and consolidation before moving to the finishing stages.

One of the reasons for the development of this process is that the capital investment required for a commercial dry forming machine is considerably less than for a conventional paper machine. This means that capacity can be gradually increased by the addition of units rather than having to make the large commitment of funds required in the installation of present systems.

Some difficulties have been encountered in the development of this process which could provide the basis for further research and development. The production of paper products with adequate strength is dependent on the bonding of the fibres which in turn is influenced by the degree of fibrillation achieved in refining. The problem in dry forming is that fibrillation can only be achieved with wet fibres, so the fibres would need to be dried after refining before being passed to the dry forming unit. Solvent de-watering using acetone has been suggested for this purpose, but no satisfactory solution has been reported.

The problem is not so critical in board manufacture. One suggestion is that unrefined dry formed sheet could be used for the middle plies in boards, where strength is not so important, with wet liners, produced in the normal way, for the outer layers.

The tear strength of dry formed paper is lower than conventional products and will not stand up to the stresses imposed in some applications, for example, cigarette packets.

Secondary fibres which have previously undergone fibrillation may be a good raw material for dry forming although there is a limit to the number of times fibres can be recirculated through a processing operation.

University research

The main opportunities for university based research appear to exist in the areas listed below.

- Improvement of refining techniques and control. The efficiency of the present refining process is low, mainly because of the difficulties encountered in effectively transferring energy to the paper fibres in order to achieve the required modifications. The use of ultrasonics may provide one approach to the solution, but there could be others.
- Investigation of the chemical preparation of secondary fibres and the effects of repeated recycling on fibre properties.
- Examination of substitute materials for wood pulp, such as straw, willow and synthetic materials.
- Further research into head box design and its influence on sheet formation. Including an examination of the effects of an increase in stock consistency on energy use.
- Improvement of mechanical drying techniques through felt development, and the enhancement of press action by pre-heating, etc.
- Further study of the combination of water with the fibres in the web and the changes that occur during drying, with particular reference to the identification of important variables and the improvement of current drying performance in the final stages.
- The development of product drying controls and monitoring systems for all stages of processing.
- Further research into dry forming techniques and the associated problems of stock preparation and final product strength.

References

Abott B A
Energy and the UK paper industry
University of Cambridge (PhD thesis), 1977

Higham R
Handbook of paper and board
The Paper and Board Manufacturers Association, 1970

Newton G E H
Private communication, 1977

Palmer A A
Some results from the Energy Audit Scheme
Paper Industries Research Association (Department of Industry Seminar)
November 1977

Webzell A B
Energy utilisation in the paper industry
Journal of the Institute of Fuel, June 1972

9 Reclamation and recycling

Chapter contents

Introduction

The central role of reclaimed and recycled materials in conserving materials and energy has been discussed, as appropriate, in each chapter so far. However, the subject is so important to the theme of this paper that the implications are more fully and generally described here.

Conservation

There is, today, a greater public awareness than ever before of the critical dependence of western civilisation on a continued supply of raw materials. During the past few years considerable literature has emerged. In such a climate it is understandable that there should be a preoccupation with recovery, but this interest has sprung largely from a popular concern over pollution rather than from a sense of finite resources and of conspicuous waste. Although the word conservation is used in this report to signify the husbandry of resources, in a wider context it is used to also embrace the preservation of the environment.

A great many works (e.g. Taylor, 1970) have been devoted to pollution and lie outside the scope of this report. These include 'The Limits to Growth' (Meadows et al, 1972), where problems of economic growth were considered under several headings, including possible shortages of resources and of cumulative pollution. The work has been widely criticised (e.g. Cole, 1973), but it has unquestionably served a useful purpose if it has focussed attention on the problem.

Definitions

The question of definitions has been debated at length but there is still much, possibly unnecessary, confusion. The following simple definitions are used here.

- Recovery or reclamation—the retrieval of materials or energy that would otherwise be discarded.
- Recycling—the act of reprocessing recovered material for use in applications for which virgin material would otherwise have been employed. Thus recovery must precede recycling, but recycling presupposes suitable purity and physical form.
- Waste or residual—material or energy from a process which it is cheaper to discard than to employ for any further purpose either in the process or elsewhere.

Recovery

Benefits of recovery

Long- and short-term materials supply

In a world where the total availability of resources is fixed (although at a practically inexhaustible level) for many materials (Callot, 1974), an ability to supply part of mankind's requirements from secondary resources will clearly extend the resource life.

In the short-term, supply from secondary sources confers a degree of immunity from market fluctuations in primary materials, whether due to price changes, shortage of capacity in the primary industry or action by some kind of cartel. Furthermore, new secondary plants can usually be built for a fraction of the capital cost of the corresponding primary installation.

Balance of payments

The advantages to a nation's balance of payments of a healthy reclamation industry are self-evident. To the materials-producing country it represents a reduced domestic demand for primary material, thus releasing more for export. To the materials-importing country it represents a saving of foreign currency, for example the value of domestic reclaimed metals and minerals from old scrap was estimated as $4 000 million in the US in 1976 (US Bureau of Mines, 1977). In the UK, the value of recycled aluminium, copper, lead, tin and zinc in 1969 was £280 million at 1969 prices.

Basis of economic comparisons

In the field of energy analysis there are at least two, conflicting, schools of thought:

- Energy analyses are independent of economic analyses; choices may be made on the basis of either or both, and this implies that any material resource can be exploited given sufficient energy. Energy should thus be treated as a unique and essential resource.
- Energy is only one of the primary natural resources that control technology.

The second of these is, initially, the more plausible since not only is it impossible to assign an energy concept of value but other primary inputs, such as capital, labour, land and minerals are also clearly indispensable (Veach Long, 1975).

The concept of efficient allocation of resources can be defined in numerous ways. It is convenient to give here those that relate to physical efficiency and economic efficiency:

- Maximum physical efficiency is attained when, through the efficient application of energy to a system and the careful husbandry of materials throughout, neither is wasted.
- Maximum economic efficiency is attained when given resources of capital, labour and all forms of natural resources are so combined that a higher output of any desired product is possible only at the cost of a reduced output of some other desired product.

Energy recovery from domestic refuse

A review of the benefits of resource recovery would be incomplete without consideration of the direct recovery of energy. Energy recovery from waste is best considered as a separate area even though most recovery systems also make more or less sophisticated attempts to recover valuable materials. As an obvious, well-documented and apparently fertile field for investigation, it will be dealt with only briefly here.

Clearly, it is advantageous to recover valuables from waste by as direct a route as possible in order to minimise the necessary investment of work. Hence, a bottle recovered from domestic refuse may have greater value as a bottle than as a source of crushed glass (cullet) provided that it may be re-used with little or no additional work, i.e. added value. If, however, it requires stringent washing, disinfection, label removal, and inspection, the necessary work may be so expensive that the value of the bottle would have been greater had it incurred the small marginal costs of bulk handling and melting. If the bottle is chipped it may be worthless for anything but the lowest grade of secondary domestic or industrial use, e.g. as a make-shift watering can.

Many organic materials cannot be reclaimed for any purpose approximating to their original use because of their potential for acquiring contamination. Thus, paper and putrescible materials in domestic refuse form a self-contaminating mixture such that even extensive processing can produce only a degraded paper product. Where the inputs necessary to achieve separation into re-usable materials may be large, it is relevant to consider applications where contamination is unimportant and separation largely unnecessary. Most of the available processes for energy recovery from domestic refuse fall into this category. Six broad classes are apparent:

- direct recovery of waste heat from incinerators;
- production of a refuse-derived fuel capable of being stored;
- use of pulverised refuse as a supplementary fuel in oil- or coal-fired boilers;
- pyrolysis, i.e. heating in closed retorts in a limited air supply to produce a mixture of gases and an oil-like liquid with significant heat value;
- hydrolysis, i.e. reaction with water to form a mixture of usable organic chemicals, notably alcohols;
- methane generation by anaerobic fermentation.

Each presents unresolved technological problems and few systems have yet generated sufficient operating data to permit a realistic economic assessment. Nevertheless, the potential for refuse-derived energy in the overall energy consumption pattern of advanced nations is substantial. In the UK, some 18 million tonnes of refuse are generated annually (Miles, 1972) and it may possess a third to a half of the heat value of coal (Schulz et al, 1975). Thus, refuse could annually contribute 6 to 9 million tonnes of coal equivalent, or 5 to 7·5% of current annual coal consumption. In the USA, annual arisings of 145 million tonnes of collected refuse (Higbie, 1974) suggest up to 70 million tonnes of coal equivalent or approximately 11% of annual domestic bituminous coal production. Savings of fossil fuels by these means would extend their life or, alternatively, free them for use as petrochemicals.

Energy savings from use of scrap metals

The use of materials such as metals that have already been refined to the purity required for their initial application may confer substantial energy savings compared with primary materials.

Total energy requirement, E (GJ/tonne), to produce copper from mine to ingot is (Chapman, February 1974)

$$E = \frac{51 \cdot 696}{G} + 27 \cdot 058$$

where G is ore grade (%).

Thus the energy requirement is approximately 79 GJ/tonne from a 1% ore and 130 GJ/tonne from an 0·5% ore. If allowance is made for the increased crushing energy likely to be necessary with progressively leaner ores, the energy requirement of extraction from a copper ore of 0·5% grade could be as high as 153 GJ/tonne. Energy requirements rise dramatically as ore grade falls and, since the progressive depletion of rich ore bodies has characterised primary copper extraction in this century, a steadily rising energy input may be expected.

Energy requirements for aluminium extraction, from bauxite mining to final ingot, have been given as about 300 GJ/tonne. It is dominated by the energy consumed in the reduction cell so total energy requirements are less dependent on the grade of ore used. The available bauxite-bearing deposits are in any event richer, by as much as two orders of magnitude, than the copper ores currently being exploited.

Energy requirements for obtaining metal from different grades of scrap must be calculated with regard to the melting losses likely to be incurred. For aluminium these can vary from 1·5% for a remelt cast shop operating on a mixed charge of ingot and heavy scrap to 2 to 10% for scalpings, up to 30% for lacquered can scrap, and up to 50% for foil. Energy requirements are summarised in Table 9·1.

Simple comparison of figures for primary and secondary production show a considerable energy saving in favour of secondary.

Table 9·1
Energy requirements for different grades of scrap (Chapman, June 1974)

Type of scrap	Energy requirement (GJ/tonne)	
	Aluminium	Copper
New scrap		
Clippings	9·3	2·2
Borings	9·8	2·6
Skimmings	9·6	21·4
Old scrap		
Sheet, etc.	9·8	21·4
Wire and cable	18·0	8·8
'Irony' scrap	14·4	22·1

Environmental

There are several environmental gains from recovery of materials. Firstly, the quantities of residuals such as mining wastes are reduced because of lower activity in the primary extractive sectors. Secondly, the collection of items whose useful life is over reduces the potential for uncontrolled disposal in the form of litter or unregulated dumps. Further, it reduces the pressure on, and extends the life of, landfill sites. Lastly, secondary benefits are available through a reduction in pollution because total energy demand of the processing industry is reduced. Whether this is realised is a function of whether the reclamation process contributes more or less pollution than does the corresponding primary extraction.

It is clear from the foregoing that recovery is, in principle, beneficial. What is now needed is a rational, quantitative analysis of the benefits and disadvantages of recovery on all the inputs and outputs of manufacturing industry, with some attempt being made to assign values to criteria related to the quality of life.

Limits of recovery

A fundamental law governing the inputs and outputs is the Second Law of Thermodynamics. This may be stated in several ways, but the most relevant to this discussion is that no heat engine, energy transformation or chemical reaction can work at 100% yield under practical conditions. Thus, 100% efficiency in any recovery process is impossible in theory as well as in practice, where accidental losses and dissipative uses of materials remove them permanently from the production cycle.

Another important thermodynamic maxim is Clausius's statement on entropy, defined either as degree of disorder or as state of chaos. According to Clausius, the entropy of the universe is increasing to a maximum, i.e. an ordered will always degenerate into a disordered state, given time. Workable deposits of metallic ore arise from their having been concentrated, by geological phenomena, into non-random pockets where they are present to many times their average crustal concentration. From these high concentrations, minerals tend once more to random distribution by natural processes such as leaching by rainfall. Man contributes to this global redistribution when metals are extracted from areas of high concentration and dissipated over the surface of the globe.

A disordered state can only be re-ordered by an input of energy. Thus, the products of organic decomposition may, ultimately, find their way into another living creature, i.e. may once again form an ordered structure. This is possible only with a continuous energy input, in this case derived from the sun. Direct solar energy will not, in general, convert the product of man's manufacturing processes into new raw materials. Normally, discarded materials can be returned to the processing stage only by an input of energy in the form of chemical or physical treatment. Moreover, energy, in one form or another, must be supplied at all stages of a multi-stage process that is not thermodynamically reversible.

The dissipation of materials reduces the possibility of their recovery because of the energy input that is required. Dilution of the costly contents of a plating bath by discharging them into a stream almost certainly means that recovery can never be a commercial proposition (although recovery may, ironically, be enforced for ecological reasons).

Economics of recovery

Economics of manufacture

It is almost axiomatic that where a non-product material can be employed economically it will be so employed, and this is true at every level, whether that of the dustman engaged in the time-honoured process of totting or of the large-scale manufacturer who may annually collect and sell to the trade many thousands of tonnes of off-cuts, borings and turnings of valuable non-ferrous metal. In both cases the motivation is profit, and if a profit cannot be made from reclamation of the material in question it will generally be abandoned. The material used for any manufacturing process will, therefore, almost always be the cheapest material of which reliable supplies are available.

The degree to which reclamation is practised will vary according to the levels of monetary costs (which are quantifiable) and social costs (which are not).

Every recycling operation takes on a monetary cost from the time that the first investment is made in labour, materials, or machinery to recover the material in question. Such extra costs are undertaken in anticipation of extra benefits. In private industry, the decision to recycle is determined by the monetary difference accruing from using virgin or primary material rather than reclaimed or secondary material, i.e. to minimise the sum, C, of the total costs of primary and secondary resources at a given output, X. The total cost can be expressed as

$$C = TC_V(X) + TC_R(X)$$

where TC_V and TC_R are, respectively, the total costs to the company of primary and secondary (recovered) resources (Pearce, 1974).

This approach, however, ignores the social costs and benefits associated with recycling. The aim of any company must be to maximise its profits. Thus, it will normally buy its raw materials as cheaply as possible and dispose of its wastes as cheaply as possible. Both of these courses place a charge on the community if they mean that no effort is made to restore the landscape to its former state and that surreptitious dumping of wastes takes place at unsuitable sites. The firm may be said to have externalised a portion of its costs, i.e. it has transferred to the community certain costs associated with amenity preservation.

The social point of view, it is argued, must aim to minimise the sum of the costs of resources as in the private case, but also accept and minimise the total external costs, including pollution, resulting from the use of virgin or recycled materials. The new total cost, S, can now be expressed as

$$S = TC_V(X) + TC_R(X) + TEC_V(X) + TEC_R(X) - B_{ERL} = L(X)$$

where TEC_V and TEC_R are respectively the total external costs of the use of primary and secondary resources,

B_{ERL} is the benefit associated with the extension of the resource life, and L(X) is the benefit associated with gains in land.

The analysis is shown in Figure 9·1. It may be seen that the private and socially optimal recycling rates do not necessarily coincide although they might do so if the recycling technologies themselves were more polluting than the disposal of wastes from a manufacturing process working on virgin materials.

Figure 9·1
Effect of recycled materials on costs of production (after Pearce, 1974)

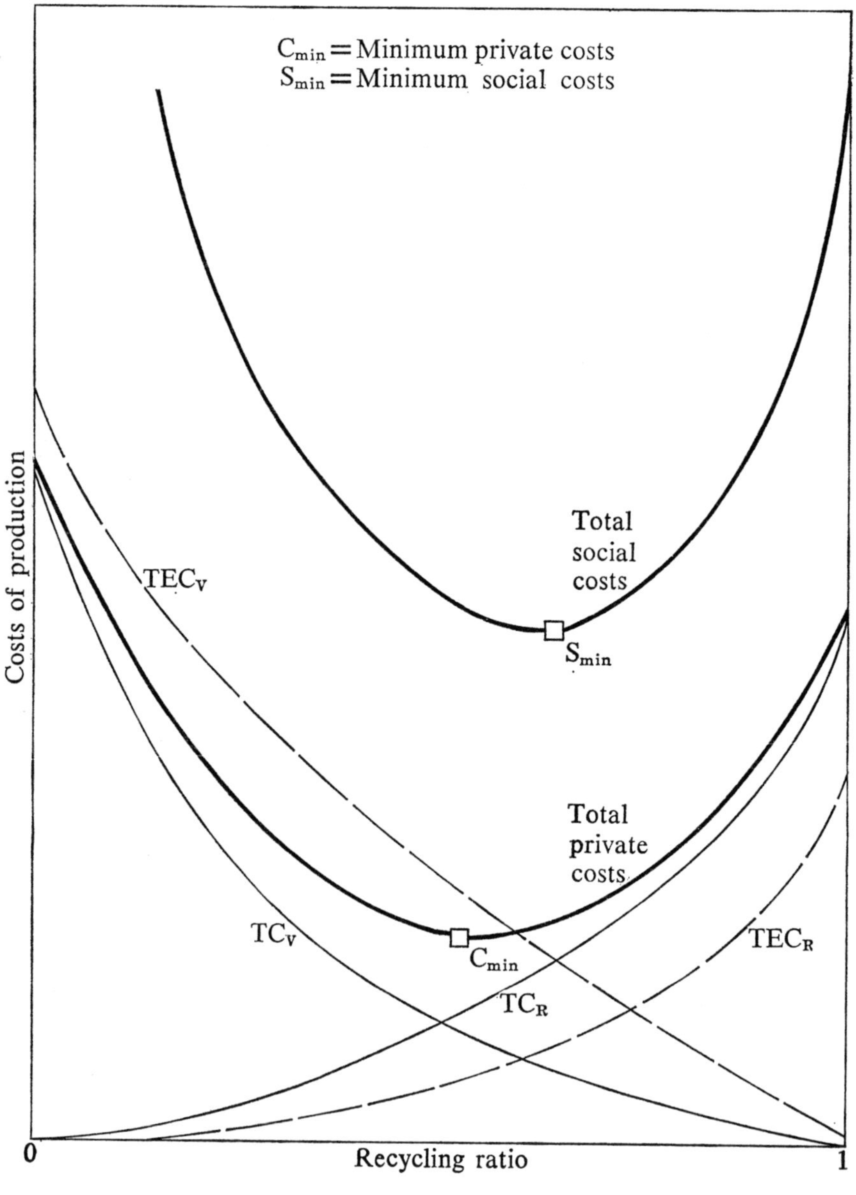

In extreme cases, on the one hand a manufacturer may seek to maximise his profits or to sell his product at the lowest possible price—and the two are not mutually exclusive—by ignoring external costs. On the other hand, a nationalised industry would presumably fix its prices to reflect such external costs. If, in order to compete with the private company, the nationalised concern has to set its prices at a lower level, it can continue trading only with external financial support, via taxation revenue.

Economics of collection

It is clearly less expensive to recover the first 10% of discarded material than the last, by virtue of its probable deposition at predesignated collection points. A further amount might easily be retrieved, at somewhat greater cost, from areas near the collection points and its monetary value might still exceed the cost of collection. The remaining material might be dispersed over so wide an area that recovery, although not impossible, would be totally uneconomic. Thus, there is a point at which the extra monetary costs of recovery outweigh the extra monetary benefits, and this is the point at which the economist would say that the optimal level of recycling has been reached (Pearce, 1974). Recovery to a level higher than this optimal level would then depend on whether the social value of avoided waste, in this case litter, and of the additional employment so created, are deemed to outweigh the resource or monetary costs of recovery, i.e. labour and fuel costs of collection, assuming throughout that there is a ready market for the collected material.

This view has been quantified in terms of the average energy requirement/tonne of recovered copper and aluminium as a function of the efficiency of recovery. Since the first material to be recovered will be the most attractive economically, the average energy requirement/tonne may be expected to rise as the less attractive arisings are brought into the net. Thus, the average energy requirement for copper may rise from approximately 2·52 GJ/tonne for the first 10% recovery to about 10·8 GJ/tonne at 100% recovery (Chapman, 1974). The analogy between energy requirement and economic desirability of scrap is not an exact one since energy costs are not yet such that all other costs may be neglected.

The social costs to the environment of some wastes, e.g. radioactive materials, might be thought so high that they should be reclaimed whatever the cost, but this is not usually the case.

It thus emerges that environmental amenities have to be purchased by society at large, either through a willingness to pay a premium for goods manufactured from (uncompetitive) recovered materials, or through the taxation necessary to finance restoration of the environment after primary extraction.

In order to establish an optimal recovery level one must take a so-called systems approach, considering all the costs and benefits associated with recovery.

The systems approach

Social aspects

The environmental and economic advantages of recovery must be considered as a part only of an entire industry or group of industries. The consequences of the implementation of greater re-use may be far-reaching, unexpected, and sometimes unwelcome. A failure to consider all the inputs and outputs may invalidate some of the conclusions of investigators whose idealism may sometimes overcome their sense of the practical.

A change in manufacture and usage patterns may lead to an immediate loss of employment in one activity that may or may not be absorbed by expansion of another. Compensation from public funds (taxation) for any unemployment so produced must be assured by the same conscious commitment that produced the shift in employment patterns. Since, however, the standards of living in western countries depend upon the prevailing degree of activity in manufacturing industry, any policy that would shift workers from productive to non-productive tasks must be implemented very carefully. In the final analysis employment has to exist for substantially all of the available workforce.

Commercial aspects

Factors influencing retail sales

A desire to save materials and lower production costs has, with increasing cost of many common raw materials, often led to complex, multi-material construction that may preclude efficient recovery of the component materials when the product is scrapped. Difficulties are sometimes a function of miniaturisation, as in the case of small pocket calculators. Until recently an office calculating machine might have weighed 10 kg, most of which was metal. Today, a vastly more versatile machine may weigh 250 g and contain very little metal, most of which is incorporated in printed circuits, where copper is almost inextricably mingled with solder, minute amounts of precious metals in the contacts, and relatively large amounts of polymeric material.

Factors influencing markets for recovered materials

Capital investment in equipment for a particular recycling operation can be justified only if a reliable market exists for the product at an economic price. On the other hand, the user of raw materials can generally rely on obtaining only a relatively small fraction of his requirements from secondary sources, simply because the amount of secondary material available at any given time is a function of the quantity of primary material going into service at the beginning of the useful life of a commodity, perhaps many years earlier. The exponential growth in materials demand then means that secondary supplies, geared to the lower demand of previous years, will always fall short of current demand. Thus, the manufacturer must rely principally on primary producers. In times of recession, a manufacturer's requirements are smaller, but since he will not wish to jeopardize post-recession supplies of primary materials by a temporary reduction in his level of orders, he is more likely to reduce his orders for secondary material.

For his part, the collector of secondary materials will find it worthwhile to collect, process and sell residuals only when market prices are buoyant. At other times he may retrench or stockpile pending a price recovery. A further complication is that secondary materials have not, in the past, always been of satisfactory quality. The situation thus arises that:

- an entrepreneur will invest in scrap-processing equipment only if he believes in the existence of a reliable market for his products at an economic price and,
- a manufacturer is only likely to want secondary materials if they are available in quantities comparable with his demands for primary materials, and if they are also of satisfactory quality.

The stabilisation of this supply-demand balance may be considered one of the more promising areas for government action.

Any material going into service may become contaminated with paint, dirt, grease or with incompatible materials termed 'pernicious contraries'. The scrap arising from these materials when recirculated tends, cumulatively, to raise the impurity level of new material stock. Since the degree of contamination that may be tolerated by materials for many applications is strictly limited, increasing contamination levels limit the proportion of scrap that may be used in the future, operating to the disadvantage of scrap producers and consumers alike.

There is, already, disturbing evidence that the levels of residuals in steel are increasing. Copper and tin contents, for example, rose markedly in UK steel around 1917, probably due to increased usage of scrap during the Great War, and have never since reverted to their former levels (Hundy, 1963). The position can only deteriorate as large quantities of, for example, tinplate scrap enter the system. It is probable that similar trends exist in those other materials whose production depends on the inclusion of significant fractions of post-consumer scrap.

The problem of increasing contamination may, in principle, be eased by prevention of indiscriminate mixing of materials at the design stage or by the development of some system of rapid identification at the scrapyard. Thus, further gross contamination of materials stocks might be avoided, although superficial contamination from grease, oil or paint would still remain. Existing stocks of heavily contaminated material, for which refining costs would exceed final monetary value, might still be used as degraded products but this would frequently require some relaxation of current specifications.

Much effort has been devoted to the development of waste segregation techniques (Henstock, 1974, 1975 and 1976). However, the initial products of such recovery operations are seldom pure, and require further energy investment for refinement. Alternatively, the systematic investigation of the properties of such contaminated products might facilitate their direct use as raw materials in so-called open-loop recycling. Here no attempt is made to regain the specification of the original virgin material. The objective is to produce a degraded product suitable for use in selected applications. Although there are many data concerning the properties of pure metals or of deliberately produced alloys, comparatively little is known about the properties of contaminated metals. Problems that are immediately foreseeable are:

- satisfactory quality may be impossible over a given period;
- manufacturers may be reluctant to use degraded materials of which they have no experience, and which might lead to unexpected failures;
- contaminated materials lack the flexibility of high purity materials and may thus be suitable only for a restricted range of uses.

The research required to establish the properties and range of uses for recycled materials is unlikely to be carried out unless there is reasonable hope of a long-term stable market. The importance of product and market stability has possibly received insufficient attention in the past, but progress is now being made to establish accurate specifications for recovered materials (e.g. Alter and Reeves, 1975).

Technological aspects

Materials could be recycled more easily if products were designed so that uncontaminated fractions could be recovered easily and at minimum total cost. In private industry this cost would be monetary, but on a national basis it might be necessary to consider other factors, such as a need to provide more employment. The concepts may be examined in the context of a major manufactured item, the car.

At present the typical dismantler removes the components that are easily marketed, normally the radiator, battery, engine and related parts, etc., and then sells the remaining shell. Unfortunately it still contains unwanted contaminants, such that it is unattractive as a source of steel scrap (US Environmental Protection Agency, 1972). The maximum acceptable level of copper in the scrap is about 0·15% and, since the steel may already contain about 0·07%, only an additional 0·08% can be left on the shell (Poliskin, 1969). Work carried out by the United States Bureau of Mines indicates that removal of the radiator and engine plus the transmission, differential, brake drums and brake cylinders still leaves a copper residue far in excess of the permissible. The copper can be removed by hand but this operation is difficult and expensive given the high labour rates prevailing in the USA. In countries where labour is plentiful and inexpensive this obstacle would not arise. The parts removed are worth very little and the only reason for removing them is to improve the quality of the light and heavy steel.

The problem could be lessened by relatively minor redesign of the car. The copper wire could, for example, be replaced by aluminium. During subsequent melting and refining of the scrap steel any residual aluminium would oxidise and enter the slag. Aluminium is harmless in steel and is, indeed, frequently deliberately added as a deoxidant at the ladle stage.

Table 9·2
The composition of a typical American car (Dean and Sterner, 1969)

Material	Fraction of total car (%)
Light steel	36·6
Heavy steel	34·2
Cast iron	14·3
Copper[1]	0·9
Zinc	1·5
Aluminium	1·4
Lead	0·6
Rubber	4·1
Glass	2·4
Other combustibles (padding etc.)	3·6
Other non-combustibles (dirt, insulation, etc.)	0·4

[1] Including copper as brass, etc. but excluding the copper in solid solution in the steel.

Copper usage could further be reduced by replacing the electric motor stators with permanent ceramic magnets.

Alternatively, reduction in copper use might be achieved by making copper-containing parts easily removable (Stone, 1968). Copper components could be located in one or two standard places such as a mounting bracket on the radiator, itself normally removed by the dismantler. The bracket could carry the horn, voltage regulator, relays for accessories such as the starter, and the solenoid. Other copper-containing components such as heater core, windscreen wiper motor, heater motor, and other relays might be attached to a plate located on the fire wall, and removal of this plate could provide easy access to the wires and instruments under the dashboard. All wires to and from these two assemblages could be consolidated in one or two wire looms, provided with pullaway connecting plugs. Body wiring could be enclosed in a plastic conduit mounted beneath the car.

Assuming that these suggested design changes were made, only about 0·04% of copper would be left which, in addition to the 0·07% contained in solid solution, would give a total content of 0·11%, i.e. well below the acceptable maximum.

Economic aspects

It must once again be emphasised that energy and other resource inputs are necessary at every stage of dismantlement, segregation and upgrading of scrap, with value changes at each stage (Alexander, 1975). For recovered materials to be attractive, the inputs must have a smaller monetary value than that needed for corresponding primary materials.

The costs are usually very well understood and easily identified for an industrial organisation, but not so for a voluntary organisation. The costs of voluntary collection are passed on by the organisation to its members, each of whom gives his time free and usually supplies his own vehicle in order to travel to the area in which the collection is to be made. The gain to the organisation may be considerable, but in a national sense the operation may have cost far more in energy, both in human nutrition and in vehicle fuel, than would have been required to extract new material from ore, even if the external costs of primary extraction are included. If the real total costs of recycling aluminium beverage cans exceeds the cost of producing new ones the exercise would need to be justified on other grounds, such as environmental improvement or occupational therapy. Alternatively, the need for a material in wartime would transcend economic factors.

University research

A number of aspects of the recovery of energy and materials which seem particularly suitable for investigation by universities, or other objective bodies funded from government sources are listed below. Whilst some could be undertaken almost independently, most would require close contact with manufacturing industry.

Technological and techno-economic

- Many proposed methods of energy recovery are at the stage when only extensive data from pilot plants would be of use. Hydrolysis of cellulose in domestic refuse (to produce sugars) and of non-metallic residues from scrap shredders is worth further scrutiny on a laboratory scale.
- Assessment of the short- and long-term effects of the build-up of contamination in materials, such as copper and tin in steel or lead and aluminium in zinc.
- Reassessment of composition specifications for materials.
- Development of rapid methods of materials identification at the scrapyard.
- Accurate measurement of melting losses and energy requirement for reclamation from different types of arising in order to determine the optimal recycling level for each type.
- Analysis of the true level of materials costs in manufacture in order to establish the true benefits of materials substitution.
- Examination of the costs of design changes to facilitate recycling of manufactured goods, for example, cars and other complex structures.
- Development of open-loop recycling where degraded products with low value are used in applications where purity is not important.

Economic and socio-economic

- Assessment of the true economics of collection (especially by volunteer groups) of paper, beverage containers, etc.
- Stabilization of supply/demand relationships for secondary materials, e.g.

paper and steel.

- Development of a systems approach so that accurate predictions can be made of all the effects of the recovery and re-use of industrial products or materials. The glass container industry is a heavy consumer of energy and might be a particularly useful subject for study.

Research recommendations from the Waste Management Advisory Council
In its first report, produced in 1976, the Waste Management Advisory Council made a number of recommendations for research, some of which are relevant to the universities. They are paraphrased below.

- Provision of more information about the volume, nature and distribution of waste.
- Assessment of the effect of reclamation on the balance of payments, taking into account indirect as well as direct costs and benefits.
- Examination of the costs of specialised collection, transport and processing as well as the combined value of the material recovered and the saving in disposal cost.
- Development of mechanical methods of sorting mixed waste.
- Increase in the proportion of waste paper used in the manufacture of lower grade paper products.
- Investigations into the recovery of tinplate waste from domestic waste and the recycling of scale, fume and other wastes arising in the steel industry itself.
- Identification of areas where recovery of non-ferrous scrap may be uneconomic but where there are considerable benefits to the country in the form of import savings.
- Increase in the reclamation of waste oils.
- Investigation of the use of wastes as a supplementary fuel in boilers.
- Improvement in the design of manufactured articles to facilitate reclamation at the end of their lives.
- Improvement in the statistics about packaging.

References

Alter H and Reeves W R
Specifications for materials recovered from municipal refuse
US Environmental Protection Agency, EPA–670/2–75–034, May 1975

Alexander W O
The recycling and disposal of solid waste
Pergamon Press (Ed. M E Henstock), pp168–180, 1975

Callot F
The problem of the world supply of ores in the year 2000 is
basically a question of relative prices
Proc. 8th World Mining Congress (Lima, Peru), Paper II.3, November 1974

Chapman P F
The energy costs of producing copper and aluminium from primary sources
Metals and Materials **8** (2) pp107–111, February 1974

Chapman P F
Energy conservation and recycling of copper and aluminium
Metals and Materials **8** (6) pp311–319, June 1974

Cole H S D et al
Thinking about the future, a critique of *The limits to growth*
Chatto and Windus for Sussex University Press, 1973

Dean K C and Steiner J W
Dismantling a typical junk automobile to produce quality scrap
US Bureau of Mines, RK 7350, 1969

Henstock M E
Modern techniques for recycling
Conservation and Recycling **1** (1) pp83–90, 1976

Henstock M E
Recycling—a nationwide concern
Resources Policy **1** (2) pp115–119, December 1974

Henstock M E
Materials recovery and recycling in the USA
Resources Policy **1** (3) pp171–175, March 1975

Higbie K B
Status of municipal refuse processing in the United States
Proc. World Conf 'Towards a plan of actions for mankind: Needs and resources—methods of forecasting' (Paris), September 1974

Hundy B B
Residual elements in carbon steels
Metallurgical developments in carbon steels
Iron and Steel Institute Special Report 81, pp75–80, 1963

Meadows D H et al
The limits to growth
Earth Island Ltd, 1972

Miles J E P
Recovery of non-ferrous metals from domestic refuse
Surveyor **140** (CXL) 4 200, pp36–38, December 1972

Pearce D W
Fiscal incentives and the economics of waste recycling: problems and limitations
Fiscal Policy and the Environment,
Institute of Fiscal Studies, 1974

Poliskin J S
Automobile steel scrap of low-residual copper
Proc. 27th Electric Furnace Conf., Metallurgical Society, American Institute of Minerals, Metallurgy and Petroleum Engineering (Detroit), pp141–142, December 1969

Schulz H, Sullivan P M and Walker F E
Characterizing combustible portions of urban refuse for potential use as fuel
US Bureau of Mines RI8044, 1975

Ralph Stone & Co. Inc.
Copper control in vehicular scrap with special emphasis on component design
US Bureau of Mines (unpublished), 1968

Taylor G R
The doomsday book
Thames and Hudson, 1970

Veach Long T (II)
Workshop report: International Federation of Institutes for Advanced Study
Workshop on energy analysis and economics
(Lidingö, Sweden), June 1975

US Environmental Protection Agency
The automobile cycle: an environmental and resource reclamation problem, 1972

US Bureau of Mines
Minerals in the US economy: ten-year supply-demand profiles for
mineral and fuel commodities (1966–1975)
Department of the Interior, 1977–703–001/69, 1977

Appendix
Government support of materials and energy conservation research – a layman's guide

Contents

Introduction

The diagram opposite, dealing as it does only with a sub-section of the Government machinery for support of research and development into industrial energy conservation, demonstrates the need for a 'layman's guide' to sources of Government funding for materials and energy conservation research. Although life for the academic research worker should be made easier by the interposition between himself and the machinery of the SRC co-ordinator, he will no doubt find it useful to consult the alphabetically-ordered list of bodies which might fund his research, both in the UK and the EEC, which follows. Many advisory committees and similar bodies which do not have their own budgets have been omitted from the list unless they have a special role in relation to funding organisations.

UK Organisations

ABRC: Advisory Board for the Research Councils

Advises the Secretary of State on civil science, particularly in the Research Councils, on support for post-graduate students and on the proper balance between international and national scientific activity. Advises on the allocation of the Science Budget. Promotes liaison between Councils and users of research.

ACORD: The Advisory Council on Research and Development for Fuel and Power

ACORD is a non-statutory body set up in 1960 as a successor to the Ministry of Fuel and Power Scientific Advisory Council. Its role is to review the R&D programmes of the Nationalised Industries and report on them to the Secretary of State for Energy. These programmes exploit and extend the established energy resources and support their effective contribution to the energy economy. The Council also advises on R&D in the fields of energy conservation and alternative energy sources. Its members are appointed by the Secretary of State and are drawn from the nationalised fuel and power industries, the oil industry, general industry and research bodies; there are also some independent members. The chairman is the Chief Scientist of the Department of Energy.

CMRB: Chemicals and Minerals Requirements Board

Requirements Boards were set up in 1973 by the Department of Industry as a means of implementing the customer-contractor principle of R&D funding put forward by the Rothschild Committee. The object of the Boards is to assess the research proposals relevant to industry received from Government research establishments, Research Associations, and industry and to recommend them for funding from the Department's research budget if they see fit. They are also responsible for the monitoring of ongoing projects and for developing research strategies for future R&D activities. Nine Boards cover the different areas of science and technology:

CMRB	Chemicals and Minerals RB
CSRB	Chief Scientist's RB
CSERB	Computers, Systems and Electronics RB
EMRB	Engineering Materials RB

Government support of research into materials and energy conservation in the materials processing industries: structure and flow of funds into universities and polytechnics

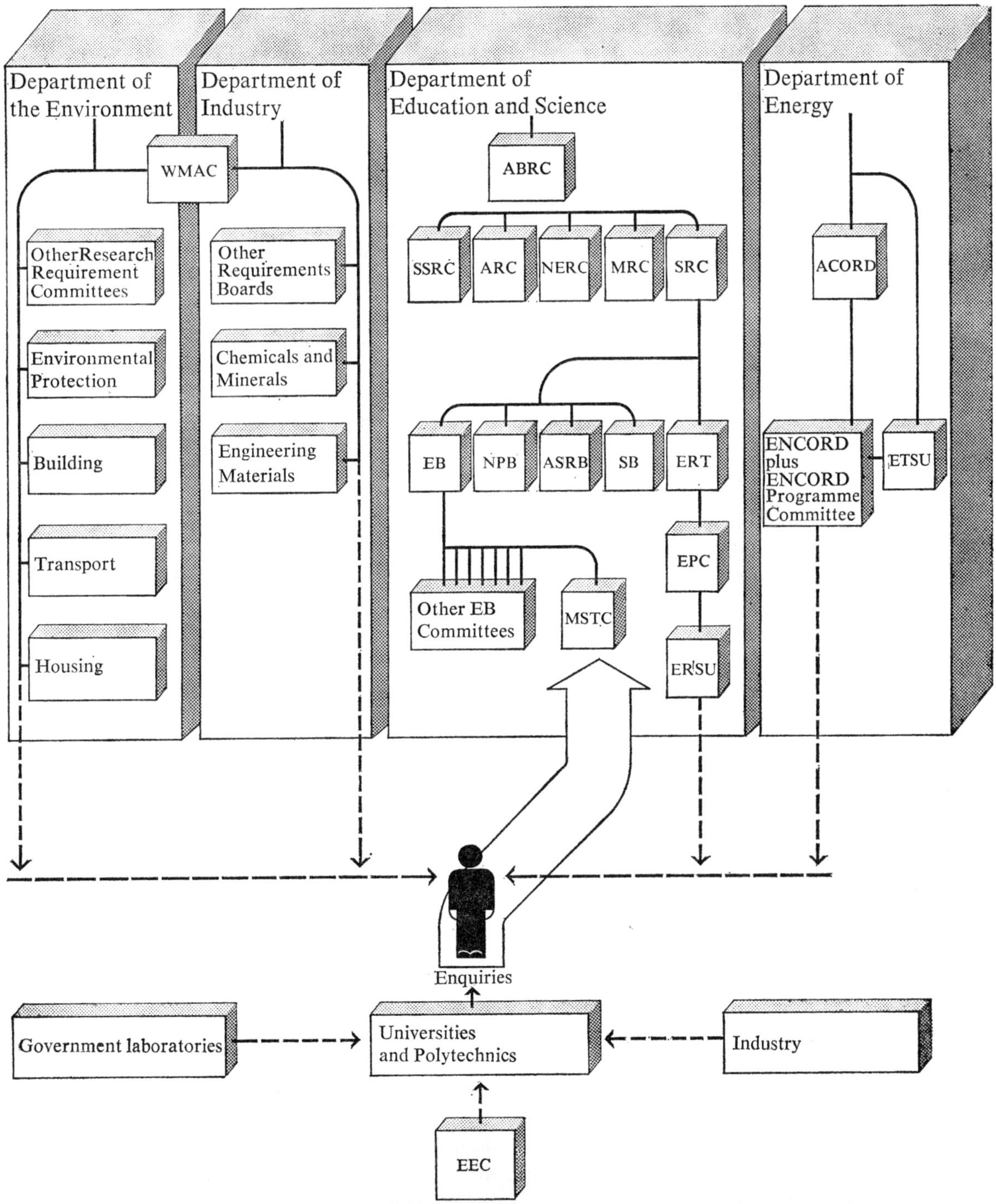

ETRB	Electrotechnology RB
GARB	Garment and Allied Industries RB
MEMTRB	Mechanical Engineering and Machine Tools RB
MSRB	Metrology and Standards RB
SMTRB	Ship and Marine Technology RB

The Chief Scientist's Requirements Board covers areas which do not naturally lie within the province of the other Boards, eg the Industrial Energy Thrift Scheme. The DoI's Energy Audit Studies were originally funded by CSRB.

EB: Engineering Board

The Science Research Council has four Boards: Astronomy, Space and Radio; Engineering; Nuclear Physics; Science. The Engineering Board administers its funds largely through its committees, to which it allocates separate budgets, but can itself fund projects, and must approve committee recommendations for individual project funding in excess of a certain value. The Board also concerns itself with policy matters: the concept of selective support of research in special areas was an initiative of the Engineering Board, and is being applied in this instance by means of the Materials and Energy Conservation programme directed by its Materials Science and Technology Committee.

EMRB: Engineering Materials Requirements Board

This DoI Requirements Board sponsors most of the research relevant to the SRC's programme in Materials and Energy Conservation in the Materials Processing Industries. It places contracts mainly with government establishments, industry and research associations rather than with universities, but an academic research worker who is prepared to collaborate with industry could put up a scheme jointly with the industry for consideration by the DoI, after consulting the SRC co-ordinator.

ENCORD: Interdepartmental Committee on Energy Conservation Research and Development

This Committee has been set up by DEn to perform the following functions:

a. Co-ordinating under the Chief Scientist's Committee the work on energy conservation R&D being funded or encouraged by different Government departments.

b. Putting forward proposals for work to be carried out by DEn.

It also submits reports on progress to ACORD.

In addition, there is an ENCORD Programme Committee which makes recommendations to ENCORD on funding of proposals for energy conservation R&D and demonstration projects. Sponsorship and management of accepted proposals are executed by teams at ETSU and NEL.

EPRRC: Environmental Protection Research Requirement Committee

The Department of the Environment runs its research programmes through the Director-General of Research, who is responsible for the general conduct of DoE's research programmes including its three research establishments (Building Research, Transport and Road Research, Hydraulics Research). Outside contracts are funded either through these research establishments or directly from the Department.

The research programmes are divided into six areas, covering Environment and Transport, and for each of these there is a research requirement committee. The full list is:

Planning, countryside, and resources
Housing
Building
Environmental Protection
Highways and Road Safety
Transport

Each Committee has a number of working groups which help to make policy and look at proposals. The DoE's Policy Division assesses the need for research in the various areas and monitors the work of the research establishments. Research proposals are not accepted direct from academia but require a sponsor, eg a research establishment (which can itself fund relevant research) or a unit at the Department itself, eg the Waste Management Advisory Council. The SRC co-ordinator should be consulted as to the most appropriate funding route.

EPC: Energy Proposals Committee
This Committee reports directly to the Science Research Council on the funding of relevant energy-related research applications, eg systems studies and multi-disciplinary proposals, which do not fall within the remit of one of the subject committees, and organises regular energy round table meetings which advise Council on policy matters. As an SRC committee, EPC is directly responsible for supporting research in universities and polytechnics in the energy area, and applications can be sent in the usual way to SRC, although prospective applicants might find it useful to consult the co-ordinator beforehand.

ERSU: Energy Research Support Unit
This is a group based at SRC's Rutherford Laboratory which carries out actions required by EPC, and prepares information papers for EPC. It also provides the EPC Secretariat. It can advise academic research workers on proposed research, and can offer support both in terms of funds and equipment or staff to university research programmes.

ERT: Energy Round Table
In December 1973, the SRC Engineering Board established a new body, the Energy Round Table, to advise it on its overall programme of energy research. At a later stage this became formally recognised as an independent advisory Council panel with the following terms of reference:

1. To review the overall balance of the Council's research programme in the energy field and to identify gaps in this programme.

2. To identify important areas for research and training requiring priority action in the energy field.

3. To consider proposals for an SRC policy on 'Energy Groups' in universities and advise Council on the same.

4. To relate the Council's programme to other national activities in this field.

The 1978 Round Table meeting was held in Oxford in September, and dealt with the conservation and efficiency-of-use aspects of energy research. The report of this meeting will be published early in 1979.

ETSU: Energy Technology Support Unit

This Unit was formed in April 1974, following discussion between DEn and the Atomic Energy Authority, and is located at Harwell. ETSU has three main functions:

1. To provide technical advisory services to DEn.

2. To carry out technical assessment studies, with particular emphasis on the elucidation of possible new areas of Government-sponsored R&D which could contribute towards the solution of long-term energy problems.

3. To manage programmes of energy R, D & D funded by Government. The technical advisory and support services for DEn include the following important items:

- Support to the Chief Scientist, DEn, and his committees.
- Acting as a focal point for DEn in the generating and collection of data in the emerging methodologies of energy accounting.
- Exploring certain educational opportunities in the energy fields.
- Providing a computer-based information retrieval system.
- Representing the Department on appropriate bodies in the UK and overseas.

MSTC: Materials Science and Technology Committee

The SRC's Engineering Board has eight subject committees reporting to it. These are:

AMEC	Aeronautical and Mechanical Engineering Committee
CEC	Control Engineering Committee
CETC	Chemical Engineering and Technology Committee
CSC	Computing Science Committee
ESEC	Electrical and Systems Engineering Committee
MSTC	Materials Science and Technology Committee
MTC	Manufacturing Technology Committee
TCEC	Transport and Civil Engineering Committee

Each committee considers university and polytechnic research applications in its area, and makes recommendations on whether they should be funded, and also concerns itself with the promotion of policy for the support of research in its area. MSTC's policy initiative led to the current programme for materials and energy conservation being accepted by the Engineering Board for special funding.

SRC: Science Research Council

The ABRC controls the budgets of five research councils:

ARC	Agricultural Research Council
MRC	Medical Research Council

NERC	Natural Environment Research Council
SRC	Science Research Council
SSRC	Social Science Research Council

The SRC funds scientific research of high timeliness and promise in universities and polytechnics, and supports post-graduate education by sponsoring courses and by funding studentships and fellowships. It funds energy research either through its Energy Proposals Committee, which reports direct to Council, or through the various subject committees of its Boards, notably the Engineering Board. The SRC co-ordinator will be happy to advise on whether a research proposal is suitable for submitting to SRC for funding.

WMAC: The Waste Management Advisory Council

This was appointed in December 1974 with the following terms of reference:

> 'to keep under review the development of waste management policies in the United Kingdom having regard to the need to secure the best use of resources, and the safe efficient disposal of wastes; to give particular consideration to ways of retaining materials from waste recycling techniques, the inter-relationship of waste utilisation and waste disposal and the reduction for transformation of waste arising; to consider the technical, economic, administrative and legal problems involved; to consider the programme of research and development; and to make recommendations.'

The Council is chaired jointly by Ministers from the Departments of Industry and Environment and has approximately 32 members. The Council has set up a number of Working Groups covering several commodity groups such as paper, packaging, ferrous and non-ferrous metals and waste as fuel. Its first report was produced in 1976.

A national anti-waste programme, NAWP, was launched in January 1976. Dr R L P Berry, a member of the WMAC, is the part-time Director who reports directly to Ministers in the Departments of Industry and Environment. His role is seen as a co-ordinator between Government agencies, local authorities, industry and the public. NAWP is serviced by the WMAC secretariat, which is administered through PTM Division of the Department of Industry.

EEC Organisations

CREST: Comite de la Recherche Scientifique et Technique

CREST was set up by an EEC Council resolution in January 1974 with the aim of simplifying contact between the Council and the Commission on matters of science and technology. Its specific task is to define objectives and to encourage a common policy in the fields of science and technology by the co-ordination of national policies and by the definition of new Community policies. It does not cover the fields of Euratom or the European Coal and Steel Community. It occupies a position between the Council of Ministers and the Commission and is the key organisation in the consideration of new research proposals. The Committee consists of representatives of member states and of the Commission. For the UK, representatives from specific Departments attend as required. CREST takes advice from the ACPM's (Advisory Councils for Programme Management), whose primary role is to advise the Commission on its R&D programme management in specific areas. Thus, the Energy Conservation ACPM deals with energy conservation R&D. Prospective applicants for EEC support should consult the SRC co-ordinator who can put them in touch with the relevant contact. The EEC also publishes details of its programmes in its Bulletin.

Directorates-General

The EEC Commission has a number of administrative sub-divisions which cover various aspects of Community affairs. Among these are the various directorates-general which are to some extent analogous to individual ministries in the UK and which have responsibilities as follows:

DG1 External Relations
DG2 Economic and Financial Affairs
DG3 Industrial and Technological Affairs
DG4 Competition
DG5 Social Affairs
DG6 Agriculture
DG7 Transport
DG8 Development Aid
DG9 Personnel and Administration
DG10 Press and Information
DG11 Internal Market
DG12 Research, Science and Education
DG13 Scientific and Technical Information and Information Management
DG14
DG15 Financial Institutions and Taxation
DG16 Regional Policy
DG17 Energy
DG18 Credit and Investments
DG19 Budgets
DG20 Financial Control

Those that have particular relevance in the context of industrial energy conservation are DG3, DG12 and DG17. Details of the current organisation of these DG's and of the rest of the EEC institutions can be found in a quarterly publication called 'Vacher's European Companion—a Diplomatic, Political and Commercial Reference Book', published by A S Kerswill Ltd.

EEC Research Projects

Research projects may be carried out in four different ways:

1. By 'Direct Action'. Projects are funded and managed by the Commission and carried out in the establishments of the Joint Research Centre of the Communities at Ispra, Geel, Karlsruhe and Petten.

2. By 'Indirect Action'. Projects are centrally managed by the Commission, but contracted out to research organisations in member states, usually with costs shared between the Commission and the contractor.

3. As a 'Complementary Programme'. If one or more member states opt out of a direct or indirect action programme the cost is shared between the remainder and this becomes a complementary programme. Because of various difficulties involved, it is not easy to get agreement for such programmes.

4. By 'Concerted Action'. These projects are co-ordinated by the Commission but the work is done in organisations of member states generally, on the basis of existing programmes—there is no financial involvement on the part of the Commission.